EEG-BASED EXPERIMENT DESIGN FOR MAJOR DEPRESSIVE DISORDER

ELSEVIER science & technology books

Companion Web Site:

http://booksite.elsevier.com/9780128174203/

EEG-Based Experiment Design for Major Depressive Disorder

AAMIR SAEED MALIK and WAJID MUMTAZ

Resources available:

- Includes links to useful reference articles, and real-world EEG data from depressed patients and healthy controls

ELSEVIER

ACADEMIC PRESS

EEG-BASED EXPERIMENT DESIGN FOR MAJOR DEPRESSIVE DISORDER

Machine Learning and Psychiatric Diagnosis

Aamir Saeed Malik
Brain9D, Adelaide, Australia

Wajid Mumtaz
New Technologies for the Information Society (NTIS), Department of Computer Science, Faculty of Applied Sciences, University of West Bohemia, Czech Republic

ELSEVIER

ACADEMIC PRESS
An imprint of Elsevier

Academic Press is an imprint of Elsevier
125 London Wall, London EC2Y 5AS, United Kingdom
525 B Street, Suite 1650, San Diego, CA 92101, United States
50 Hampshire Street, 5th Floor, Cambridge, MA 02139, United States
The Boulevard, Langford Lane, Kidlington, Oxford OX5 1GB, United Kingdom

British Library Cataloguing-in-Publication Data
A catalogue record for this book is available from the British Library

Library of Congress Cataloging-in-Publication Data
A catalog record for this book is available from the Library of Congress

ISBN: 978-0-12-817420-3

For Information on all Academic Press publications
visit our website at https://www.elsevier.com/books-and-journals

Publisher: Nikki Levy
Acquisition Editor: Natalie Farra
Editorial Project Manager: Kristi Anderson
Production Project Manager: Omer Mukthar
Cover Designer: Mark Rogers

Typeset by MPS Limited, Chennai, India

Dedication

Dedicated to

Umama and Reham (from Aamir Malik)

To my parents, Dr. Saadia Irfan and Abdul Hadi (from Wajid Mumtaz)

Contents

For additional information on the topics covered in the book, visit the companion site: https://www.elsevier.com/books-and-journals/book-companion/9780128174203

About the Authors

Dr. Malik has a BS in Electrical Engineering from the University of Engineering and Technology, Lahore, Pakistan, an MS in Nuclear Engineering from the Quaid-i-Azam University, Islamabad, Pakistan, and a second MS in Information and Communication and a PhD in Information and Mechatronics from the Gwangju Institute of Science and Technology, Gwangju, Korea. He has more than 20 years of research experience and has worked for IBM, Hamdard University, the Pakistan Government, Yeungnam University and Hanyang University in Korea, and Universiti Teknologi PETRONAS in Malaysia. Currently, he is a consultant in neurosignal and neuroimaging in Australia. He is a fellow of IET and a senior member of IEEE. He is board member of Asia Pacific Neurofeedback Association (APNA) and a member of the Malaysia Society of Neuroscience (MSN). His research interests include neurosignal and neuroimage processing and neuroscience big data analytics. He is the author of four books and a number of international journal and conference papers with more than 2000 citations and a cumulative impact factor of more than 200. He has a number of patents and copyrights and has received numerous awards.

Dr. Wajid Mumtaz completed his PhD from the Center for Intelligent Signal and Imaging Research (CISIR), Universiti Teknologi PETRONAS (UTP), Malaysia in 2017. He continued as postdoctorate researcher from 2017 to 2018 at the same institution. Recently, he joined the University of West Bohemia, Pilsen, Czech Republic, as a Postdoctoral Research Fellow. He received his MS in computer engineering and his bachelor's degree in electrical engineering from the University of Engineering and Technology, Taxila, Pakistan, in 2009 and 2005, respectively. His research interests include biomedical signal processing and applications, machine learning application to medical problem-solving such as diagnosis and treatment assessment, as well as adaptive noise cancellation for real-world data such as electroencephalogram (EEG).

Preface

This book provides a multidisciplinary resource for major depressive disorder (MDD) also simply termed as depression. The main theme of this book is to discuss the details of electroencephalography (EEG)-based methods that can provide solutions to the diagnosis and treatment-related issues for depression. The book addresses the depression-related challenges by studying both neurobiological and electrophysiological point of views. *The book* shows that MDD is a mental illness and needs proper diagnosis and patient care. Hence, the authors believe that this book can become a resource to increase awareness against depression. In addition, the book may benefit clinical practitioners, researchers, and depressed patients.

The book introduces new treatment strategies for clinicians. According to existing practice, the diagnosis of depression involves structured clinical questionnaires such as the Beck Depression Inventory (BDI) and Hospital anxiety and depression scale (HADS). However, these questionnaires have a subjective nature and could be inefficient in particular cases. Therefore, the investigation of objective methods involving EEG technology could open new treatment strategies for depression. Ultimately, improvement in the quality of life for depressed patient is possible.

From the perspective of researchers, the book furnishes information on an EEG-based experiment design involving depression. Thanks to the digital age, the advancement in the implementation of machine learning techniques has opened the door for ubiquitous applications, and depression is not an exception. However, the idea is still in its infancy, as the clinical application of EEG-based machine learning methods for depression still need research efforts to achieve clinical validation. Therefore, this book provides examples of EEG-based methods for depression termed Intelligent Treatment management system (ITMS) for depression. The ITMS exhibits improved machine learning solutions and shows clinical relevance as well.

Chapter 1, Introduction: Depression and Challenges, introduces depression and its subtypes and associated challenges. Moreover, the chapter emphasizes the development of objective methods for diagnosis and treatment efficacy assessments involving depression. In particular, this chapter shows examples of the EEG research studies that have addressed the research area on EEG-based methods for depression diagnosis and prognosis.

Chapter 2, Electroencephalography Fundamentals, elaborates on the fundamentals of electroencephalography, including EEG frequency bands, EEG recording techniques, EEG references, EEG artifacts, and methods for EEG-based artifact reductions. The EEG artifact reduction section explains the offline and online methods developed for multiple channels and single channel EEG data. The references for each technique provide guidance for detailed information on each method.

Chapter 3, Electroencephalography-Based Brain Functional Connectivity and Clinical Implications, discusses the EEG-based functional connectivity (FC) methods and their clinical implications. In addition, the chapter tabulates various applications of the EEG-based functional connectivity methods for mental illnesses such as MDD, epilepsy, schizophrenia, alcohol-use disorder (AUD), and Alzheimer's disease. Furthermore, this chapter furnishes a list of open-source MATLAB-based toolboxes focused on developing EEG-based FC methods.

Chapter 4, Pathophysiology of Depression, discusses the pathophysiology of depression, especially detailing possible causes for depression based on evidence provided by animal and human studies. The chapter emphasizes fusing of information from different sources, including evidence from human and animal studies and the functional level information from the neuroimaging modalities. The underlying causes and effects of depression include brain volume losses involving different brain areas.

Chapter 5, Using Electroencephalography for Diagnosing and Treating Depression, reviews research work involving EEG-based methods for diagnosis and treatment efficacy assessment for depression. The chapter provides details on the EEG and event-related potential (ERP) methods. The chapter explains common findings on the fusion of neurobiological and electrophysiological data. In addition, important research gaps based on this information are highlighted.

Chapter 6, Neural Circuits and Electroencephalography-Based Neurobiology for Depression, focuses on two main areas of study: (1) the neural circuits implicated during depression and (2) the neurobiology of EEG/ERP-based markers for depression. The first section explains the underlying neural circuits, including important brain areas implicated during depression. The purpose of the second section is to discuss theories regarding the underlying mechanisms defining EEG biomarkers for depression.

Chapter 7, Design of an Electroencephalography Experiment for Assessing Major Depressive Disorder, provides a detailed design of the EEG-based experiment involving issues such as the sample size calculations, study's exclusion and inclusion criterion, experimental setup for EEG/ERP data recording, and the selection of clinical questionnaires for depression. Moreover, the chapter provides information of the study participants in tabular form, including their clinical scores regarding disease severity, demographic information, and the total number of participants included in the study based on the sample size formula.

Chapters 8 and 9 provide a machine learning (ML) scheme, termed *Intelligent Treatment Management System (ITMS)* for unipolar MDD patients. Chapter 8, Electroencephalography-Based Diagnosis of Depression, provides details of the EEG-based diagnosis for depression (*ITMS diagnosis*) by classifying MDD patients and healthy controls. Chapter 9, Electroencephalography-Based Treatment Efficacy Assessment Involving Depression, explains EEG-based prediction of treatment outcome of antidepressant therapy (*ITMS treatment selection*), while classifying the MDD patients as either treatment respondents (R) or nonrespondents (NR) based on the pretreatment EEG data acquired from the MDD patients. The EEG-based scheme presented in these chapters inherently involves subprocesses such as noise removal from the EEG data, EEG-based feature extraction, feature selection, classification, and validation, including 10-fold cross-validation (10-CV). This chapter provides the technical details regarding these subprocesses.

All the chapters have summary and references at the end of each chapter. The readers can use the references to get more detailed information on a particular topic. We hope that readers will enjoy this book and that the information provided will be useful for their practices and research.

Acknowledgments

The task of experimental data collection was made possible with the support of our collaborators at Hospital Universiti Sains Malaysia (HUSM). Especially, we would like to acknowledge the hospital's senior psychiatrist Dr. Mohd Azhar Mohd Yasin who provided guidance during ethical approval from the ethics committee as well as assisting in experiment design, patient recruitment, and access to the office for EEG data collection. We would also like to acknowledge the logistic support and guidance provided by Dr. Faruque Reza at Hospital Universiti Sains Malaysia (HUSM). In addition, Ms. Nurul Aida Bt. Yaqoob assisted in recruiting and managing the female study participants and during miscellaneous clinical procedures.

We also thank our colleagues for the many fruitful research discussions. We would especially like to thank Dr. Hafeez Ullah Amin, Dr. Likun Xia, Dr. Ibrahima Faye, Dr. Nidal Kamel, and Dr. Naufal B. Saad for their stimulating debates.

CHAPTER

1

Introduction: Depression and Challenges

1.1 INTRODUCTION

Major depression, also termed as major depressive disorder (MDD), unipolar depression, clinical depression, or even simply depression, is a mental illness. According to the World Health Organization (WHO), depression has been identified as a leading cause of functional disability, worldwide. About 300 million people have been reported suffering from depression, globally.[1] In addition to the functional disability caused by depression, it may lead to suicide ideations. Moreover, the treatment management for depression has been challenging due to multiple factors, such as the suitable selection of medication for a patient being based on the subjective experience of clinicians and which might not be appropriate for the patient and could result into unsuccessful treatment trials. Another implication is that the patient may stop the treatment.

In this chapter, the topics covered in this book are introduced by providing a basic explanation of the relevant concepts which will be elaborated on in later chapters. More specifically, this chapter explores the possibilities of utilizing electroencephalogram (EEG) as an objective method for the diagnosis and treatment efficacy assessment for depression. Also, depression will be discussed from different perspectives such as its subtypes, signs and symptoms, the challenges associated with treating depression, an overview of the literature involving EEG studies for depression, EEG as a modality, and the basics of an EEG-based machine learning (ML) approach.

EEG-based diagnosis of depression may be compared with the conventional practice of treating depression. Conventionally, depression has been diagnosed according to criteria in the Diagnostic and Statistical Manual (DSM)-V and its earlier versions. The DSM-V provides a questionnaire-based assessment that depends on the patient's

DOI: https://doi.org/10.1016/B978-0-12-817420-3.00001-1

feedback. However, misreporting may occur when patients do not explain their condition well. Therefore an objective assessment provided by EEG may assist clinicians during clinical decision making. In addition, EEG-based methods may help standardize clinical decision making for depression.

1.2 DEPRESSION AND SUBTYPES

Several different types of major depression have been identified, for example, unipolar depression, bipolar disorder (or manic depression), dysthymia, postpartum depression, atypical depression, psychotic depression, and seasonal effective disorder.[2] *Major or unipolar depression* is the most generic form of depression. It has been characterized based on a depressed episode that persists for at least 2 weeks rendering the patient's functionally disabled. Moreover, it has been discovered as a leading cause of disease burden for women in high-, middle-, and low-income countries.[3] In the United States, it has been declared as the most common cause of functional disability.[4] For example, the prevalence of unipolar depression has been found in 13%–16% of the total US population.

Bipolar depression normally manifests as two different episodes: a depressive episode and a manic episode. The occurrence of manic episodes differentiates *bipolar* from *unipolar depression*. However, bipolar depression is less common than unipolar depression. According to National Institute of Mental Health (NIMH), it has affected 2%–3% of the americal adult population. (https://www.nimh.nih.gov/health/statistics/bipolar-disorder.shtml). Other forms of depression such as *postpartum depression* is a form of depression that affects 5% of women in their second half of menstrual cycle, 10% of pregnant women, and 16% of women 3 months after giving birth.

Some other forms of depression which are normally considered to be less common include *psychotic depression*, *atypical depression*, *seasonal effective disorder*, and *dysthymia*. *Psychotic depression* has been characterized as a mental state with false beliefs (delusions) or false sights or sounds (hallucinations). It is a more severe form of depression, but is less common as about 20% of depressed patients may have psychotic symptoms. Similarly, *atypical depression* and *seasonal effective disorder* are forms of depression that occur only during specific seasons, particularly, winter. *Dysthymia*, which may entail less severe but longer lasting symptoms than of depression, has been found in only approximately 1.5% of adult Americans (https://www.nimh.nih.gov/health/statistics/persistent-depressive-disorder-dysthymic-disorder.shtml). As unipolar depression is the most common and affects the largest population, this book mainly focuses on the patients with *unipolar depression*; all other forms of depression are out of scope. In this book, *unipolar depression* is termed as MDD, or simply as depression.

1.3 SIGNS AND SYMPTOMS OF DEPRESSION

The two core symptoms of depression are *low mood* and *lack of pleasure* from pleasurable activities. Depression involves sad episodes that prevail for more than 2 weeks and renders the patient functionally disabled. On the contrary, normal sadness that may result because of routine matters or a social problem may not be considered as depression, which is recurrent and comorbid in nature. Hence, depression should be treated properly by a specialist such as a psychiatrist or psychologist. Other symptoms of depression include:

- significant changes in appetite or weight;
- insomnia or hypersomnia nearly every day;
- psychomotor agitation and retardation;
- fatigue or loss of energy almost every day;
- feeling of uselessness or inappropriate guilt;
- decreased ability to think, concentrate, or to make decisions nearly every day; and
- recurrent thoughts of death or suicidal ideas, plans, or attempts.

1.4 UNIPOLAR DEPRESSION AND CHALLENGES

The treatment management for depression has been associated with two serious issues. First, a successful *diagnosis* of depression is required during a patient's care. Since MDD is heterogeneous and comorbid in nature, there is a high chance that MDD patients may be misdiagnosed as having a bipolar disorder during their first visit to a psychiatric clinic.[5] Because of such a misdiagnosis, the appropriate treatment process could be delayed. In addition, patients could be mistreated involving unsuitable medication (antidepressants) that may further complicate the patient's condition, for example, development of a treatment resistant scenario. Hence, an accurate *diagnosis* increases the chances to achieve remission (absence of symptoms) early. Currently, the diagnosis of depression involves the use of well-structured questionnaires such as those provided in the DSM-V.[6] Since the questionnaires are subjective in nature as the feedback from the depressed patients is required, there is a probability that the patients may not reveal their true conditions.

Hence, the reliability of questionnaire-based diagnosis depends on the expertise of the specialist handling the patient. For example, in some cases a patient initially diagnosed with unipolar depression may revert from unipolar depression to psychotic depression after 2 weeks of treatment.

Second, prediction of the treatment outcome of antidepressant therapy for a depressed patient has been challenging, termed here as the antidepressant's *treatment efficacy assessment or treatment selection*. A successful prediction could lead to a suitable selection of antidepressants for the MDD patient. Currently, the selection is subjective and mainly based on clinical expertise including an analysis of the patient's symptoms and medical history. Unfortunately, a nonresponse to an antidepressant could be a waste of the adequate time frame of 2–4 weeks and may lead to a second-time selection. Eventually, the selection of antidepressants might become a sequential iterative treatment process.[7] Hence, the inappropriate selection of antidepressants would result in poor quality-of-life for those MDD patients. In extreme cases, MDD patients may even abandon the treatment. Hence, to improve the antidepressant's *treatment selection*, an early and effective treatment strategy is required.

The treatment for MDD patients includes multiple therapies such as psychotherapy, pharmacotherapy, electroconvulsive therapy, and neurofeedback, or a suitable combination designed by an expert, for example, a psychiatrist. Specifically, pharmacotherapy involves the administration of antidepressants and is usually applied when psychotherapy alone is believed to be noneffective. The administration of antidepressants has been considered as the first-line treatment for depression involving the selective serotonin inhibitors (SSRIs), a class of antidepressants.[8] SSRIs include more than 20 different antidepressants with similar mechanisms-of-action (MOAs) that are available commercially. Antidepressants have been associated with low treatment efficacy due to treatment nonresponse as initial treatments often do not lead to recovery.[9] For example, according to a study on Sequenced Treatment Alternative to Relieve Depression (STAR*D), MDD patients achieved modest rates of remission with first treatment. For example, according to a study on sequenced treatment alternative to relieve depression (STAR*D), MDD patients achieved modest rates of remission with first treatment, that is, 47% which was even less than half the total study participants.[10] In addition, only a few patients who received adequate pharmacotherapy could actually achieve remission defined as the absence or near absence of symptoms.[10] Hence, MDD has been associated with functional disabilities, low treatment efficacy, high social burden, and medical costs.

1.5 ELECTROENCEPHALOGRAPHY AS A CLINICAL MODALITY

The use of EEG has been considered as a standard clinical modality because it offers a noninvasive and a low-cost solution to various psychiatric conditions. For example, EEG is suitable for applications such as diagnosing and predicting the occurrences of epileptic seizures,[11,12] ancillary evidence of brain death[13], quantifying sleep stages,[14] and indexing for anesthesia monitoring.[15] More specifically, EEG has been utilized for the diagnosis of depression[16,17] as well as the prediction of antidepressant's treatment response.[18,19] The digital version of EEG, known as quantitative EEG (QEEG), provides computer-based solutions to solve complex real-world issues. A detailed description involving the EEG-based methods for depression shall be provided in later Chapters 5, 8 and 9.

In psychiatry, EEG/event-related potentials (ERP) data could be utilized for two main applications: (1) as a diagnostic tool to discriminate MDD patients from healthy controls among a study population and (2) as biomarkers to generate scientific evidence of treatment outcome involving antidepressants, also termed as antidepressant's *treatment efficacy assessment*.[20,21] QEEG includes various time and frequency domain techniques referred to as digital signal processing and computational psychiatry.[22–24] In the context of MDD, various research studies have extracted information from EEG/ERP data to develop EEG/ERP-based methods for the treatment management of MDD.[25,26] The EEG/ERP-based methods have shown promise as biomarkers for treatment selection. Furthermore, such methods help in identifying patients who can continue current treatment. Hence, the treatment selection is improved by the justification of the suitability of an antidepressant and helps clinicians provide patient care. In addition, successful prediction effectively avoids the possibility of time-consuming treatment trials and improves the patient's quality-of-life.

1.6 ELECTROENCEPHALOGRAPHY-BASED MACHINE LEARNING METHODS FOR DEPRESSION

ML has found its place in the diagnosis and treatment efficacy assessment for mental illnesses such as diagnosing a depression state[16] or the automatic identification of epileptic EEG signals.[11] ML is a form of artificial intelligence as a ML model can learn from the features or input data in the form of variables. The learned classification model can be used to classify a completely new data set within the learned space of

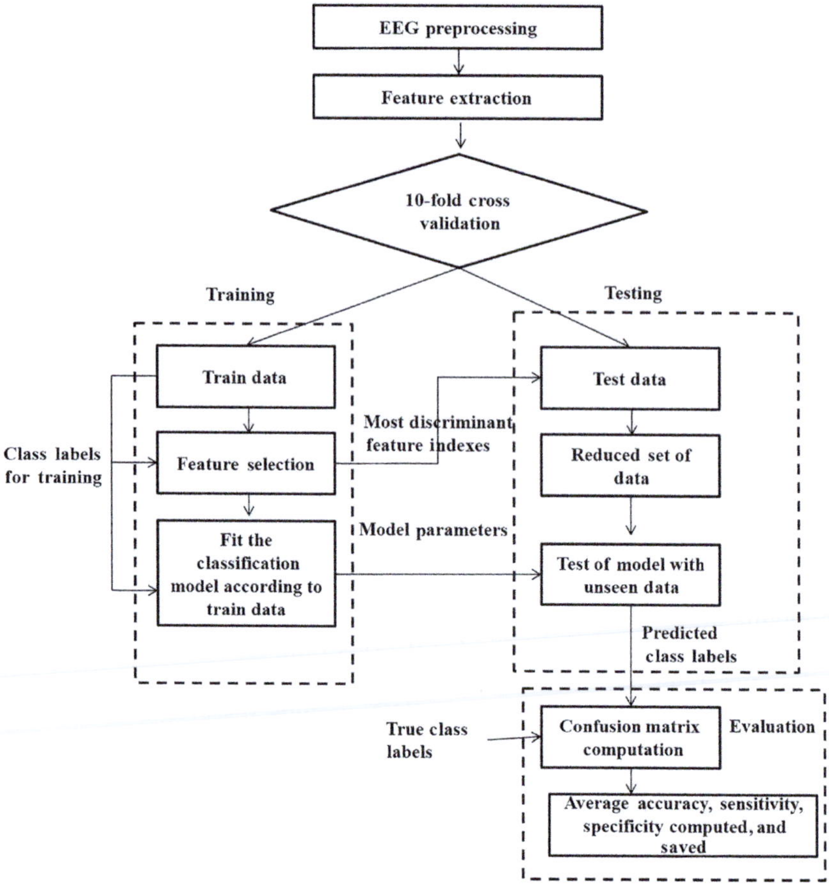

FIGURE 1.1 A general ML scheme for EEG data analysis.

the classifier. For example, a trained ML model naïve to a completely new set of data can classify it correctly. A simple example of a ML scheme is shown in Fig. 1.1 including the basic structure of a ML model such as data preprocessing, feature extraction, selection of the most suitable features, and the training and testing of the classification model. Usually, the validation of the ML model is performed with *k*-fold cross validation. A brief description on each of these subprocesses is provided next.

1.6.1 Data Preprocessing

Data preprocessing implies the removal of noise due to the rejection or correction of the artifacts from the data. An artifact is an irrelevant additive signal found in the EEG data that may suppress the actual neural activity. Due to such artifacts, the EEG analysis could be confounded and misleading. Common types of EEG artifacts include eye-blinks, eye-movements, muscular and heart activities, etc. Eye-blink artifacts are higher in amplitude than the normal EEG signal. A detailed description of the EEG artifacts and artifact removal or cleaning methods is provided in Chapter 2, EEG Fundamentals.

1.6.2 Feature Extraction

After successful artifact cleaning, the EEG data can be subjected to various kinds of feature extractions including time-based methods, frequency-based methods, and time–frequency-based methods. Examples of frequency-based methods involve the EEG signal power extraction in different frequency bands; the time-based methods could be EEG sample entropy, approximate entropy, and fractal dimension calculations; and the time–frequency methods may include the wavelet decomposition of the data. A detailed description of the EEG-based feature extraction method is provided in the following Chapters 5, 8 and 9.

1.6.3 Feature Selection

Tremendous advancements in the digital signal processing domain have introduced various new feature extraction methods. As an implication, various features can be extracted from the data. However, the features could be redundant or irrelevant when compared with the ground truths of the problem at hand. Therefore the feature selection has profound importance. The feature selection is an ongoing research area. The feature selection result into a reduced set of most key features that can be utilized for classification purposes. The features selection is important because it can significantly improve classifier performance because overfitting or underfitting problems can be avoided.

1.6.4 Classification

A classifier could be a statistical model that provides a mathematical function between the input features and the output target vector. A classification model requires proper training and testing based on the recorded data. The data can be the extracted features arranged in the form of a matrix, termed as a data matrix. The rows of the matrix correspond to the samples while the columns represent the number of features. A high-dimensional data matrix entails a substantial number of features. Usually, the data matrix is subjected to features selection to improve the classification performance as the redundant and irrelevant features might overfit or underfit a classification model.

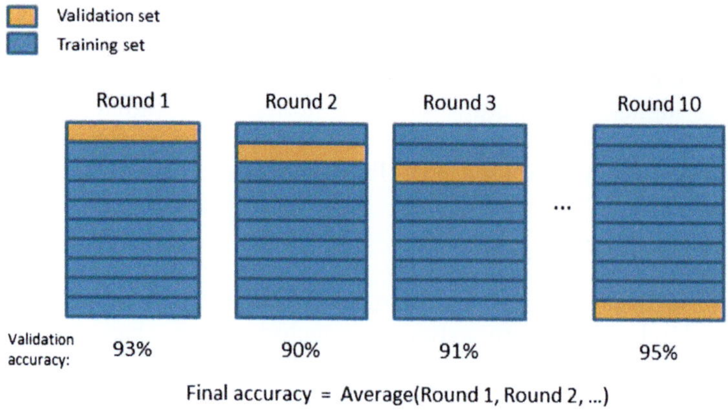

FIGURE 1.2 Ten-fold cross validation.

Classification can be divided into two broad categories, supervised and unsupervised classification. Supervised classification requires information, such as the ground truths, for training purposes. For example, the support vector machine (SVM) is a supervised classifier and requires information on the target vector that can serve as the ground truth and can be obtained during data collection. On the contrary, unsupervised classification such as the *k*-means clustering is independent from the availability of the target vectors.

For any classification model, validation is important to quantify the classification performance. For this purpose, methods such as the *k*-fold cross validation can serve better. The k value can be any value, although the most commonly employed values are 10, 5, etc.

1.6.5 The 10-Fold Cross Validation

Fig. 1.2 shows an example of a 10-fold cross validation method. The samples in a data matrix are randomly distributed among the training and validation (testing) samples. The objective of doing this is to ensure that each sample of the data matrix is utilized as a training and testing sample. This allows the calculation for the classifier accuracy, sensitivity, specificity, and other performance metrics.

1.7 ELECTROENCEPHALOGRAPHY-BASED DIAGNOSIS FOR DEPRESSION

Various EEG-based measures or features have shown an association with depression when compared with a healthy control group. For validation of the differences between the two groups under observation, it is important to employ statistical methods and for generalization purposes a classification model can be trained. In ML terms, the generalization means how a trained ML model can perform or classify a completely naïve data set within its problem domain. This could be achieved only when a properly trained ML model has learned specific patterns in the data set that can significantly discriminate the study samples into respective groups. Hence, the core of a properly trained model is the identification of a specific pattern and generalization.

A trained ML model can automatically classify new data or samples. This automation of the method enables easy classification of the depressed patients and health controls, termed as the diagnosis of depression. Also, a trained ML model can classify whether a patient can be a treatment responder or nonresponder, termed as the prediction of treatment efficacy assessment. These objectives can be achieved based on the ML concepts where the relevant features can be used to train the classification model. The trained model can then perform the desired classification. The literature involving the EEG-based diagnosis of depression has recorded many different techniques. Table 1.1 lists a summary of these methods.

ML techniques have received considerable attention due to their capability to mine (referring to data mining) noninvasive neuroimaging data to establish the CAD-based solutions that facilitate the diagnosis process.[25,26,35] For example, mining functional magnetic resonance

TABLE 1.1 The EEG-Based Methods for the Diagnosis of Depression

Study	Reference/ Number of EEG Electrodes	EEG-Based Features and Classifier	Classification Accuracy
27	Left and right mastoids/33 electrodes	EEG feature extractor called kernel eigen-filter-bank common spatial pattern (KEFB-CSP), support vector machine (SVM) classifier	~80% classification accuracy
28	REST/19 electrodes	Power of different EEG frequency bands and EEG alpha interhemispheric asymmetry, logistic regression (LR), SVM, and naïve Bayesian (NB)	SVM (accuracy = 98.4%, sensitivity = 96.66%, specificity = 100%)
29	REST/19 electrodes	Synchronization likelihood (SL) features LR, SVM, and NB	SVM classification accuracy = 98%, sensitivity = 99.9%, specificity = 95% and *f*-measure = 0.97

(Continued)

TABLE 1.1 (Continued)

Study	Reference/ Number of EEG Electrodes	EEG-Based Features and Classifier	Classification Accuracy
16	NA	Chaos theory and nonlinear dynamic methods	–
30	Bipolar/Fp1-T3 and Fp2-T4/2 electrodes	Nonlinear measures of approximate entropy (ApEn), sample entropy (SampEn), renyi entropy (REN), and bispectral phase entropy (Ph) probabilistic neural network (PNN), SVM, decision tree (DT), *k*-nearest neighbor algorithm (*k*-NN), NB classification, Gaussian mixture model (GMM), and Fuzzy Sugeno Classifier (FSC)	Classification accuracy of 99.5%
31	Bipolar/Fp1-T3 and Fp2-T4/2 electrodes	Relative wavelet energy (RWE) and artificial feedforward neural network	Classification accuracy and its value of 98.11%
32	Bipolar/Fp1-T3 and Fp2-T4/2 electrodes	SVM classifier	An average accuracy of about 98%, sensitivity of about 97%, and specificity of about 98.5%
33	NA/19 electrodes	Power of four EEG bands and four nonlinear features including detrended fluctuation analysis, Higuchi fractal, correlation dimension, and Lyapunov exponent *k*-nearest neighbor, linear discriminant analysis, and LR	Classification accuracy of 83.3%
34	LE reference/21 electrodes	EEG power, frequency, asymmetry, and coherence Linear discriminant analysis	Classified 91.3%

imaging (fMRI) data with ML methods has shown promising research results.[36,37] Specifically, the SVM is emphasized as a method of choice for the diagnosis of depression.[38] Moreover, the automated EEG-based ML methods proved feasible to discriminate the depressed patients from healthy controls.[16,32,33,39,40] In addition to depression, the classification algorithms are also found useful for neurological diseases such as schizophrenia and Alzheimer's disease.[41] In the context of depression, a classifier such as artificial neural networks can be trained to classify depressed and healthy controls.[31,42] Recently, a depression diagnostic index was proposed based on nonlinear features extracted from EEG data.[32] In contrast to fMRI, EEG offers high temporal resolution and comparatively low cost which makes it suitable for portable and remote clinical applications, especially for monitoring epileptic patients,[11,12] quantifying different sleep stages,[14] and indexing for anesthesia monitoring,[15] etc. Furthermore, EEG data acquisition is faster than fMRI and trained nursing staff (rather than only expert psychiatrists) can handle the EEG-based CAD system.

More recently, a further advanced version of an ML method termed as deep learning method has been proposed for the automatic diagnosis of depression.[17] The authors have claimed that instead of having a feature extraction and a feature selection stage, the EEG data can be directly utilized as input data to classify depressed patients from healthy controls. The automatic identification of the most relevant features is possible because of the multiple layers of a deep neural networks. The deep learning model involves multiple layers of the convolutional neural network, the softmax, and an output layer.[17] The authors have reported a classification accuracy of more than 83%. This method can be extended with different number and combination of layers in the network.

In the literature, various nonlinear features such as detrended fluctuation analysis, Higuchi fractal, correlation dimension, and Lyapunov exponent are extracted from EEG signals and have shown promise for MDD diagnosis, for example, a recently performed study achieved 90% accuracy for discriminating MDD patients and healthy controls.[33] MDD has been associated with cognitive deficits and functional impairments involving the frontal and temporal cortexs[43,44] According to a recent review, abnormalities such as MDD tend to exhibit decreased left frontal brain activity (measured as increased interhemispheric alpha power/amplitude).[45] In another study, greater left frontal brain activity is associated with less depressive symptoms.[46] In addition, EEG interhemispheric alpha asymmetry is concluded to be a risk marker for MDD because of the finding that study participants with lifetime depressive symptoms have shown less relative frontal cortex activity compared to subjects without depression.[47]

The importance of interhemispheric alpha asymmetry in the diagnosis of depression is evident from numerous studies.[48–50] For example, psychomotor retardation during depression is linked with interhemispheric alpha asymmetry[48]; EEG frontal asymmetry is considered as a marker for vulnerability of depression[49]; and decreased interhemispheric alpha waves are observed during depression.[50] In addition, the altered structure of EEG oscillatory pattern is reported in MDD patients.[45] Hence, this evidences adds to the confidence on the EEG interhemispheric alpha asymmetry as a feature to be used for automatic diagnosis of depression.

In addition to the alpha band, activity in other bands such as the theta band has shown relevance, such as a decreased frontal theta activity has also been reported.[34,51,52] Moreover, hypoactivation of the left frontal brain region[53,54] and hyperactivation in the right frontal region[51] have been reported during MDD. Despite all research findings, the clinical applications of the frontal interhemispheric alpha asymmetry and frontal midline theta band have been largely unclear.[55] Hence, further investigation on the applications of EEG features, such as EEG interhemispheric alpha asymmetry and spectral power of different EEG frequency bands, is required for diagnosing MDD.

Recently, neuroimaging modalities such as EEG signals have been successfully utilized to diagnose and predict an antidepressant's treatment efficacy for MDD.[25,26,35] For this purpose, EEG-based features such as the EEG alpha interhemispheric asymmetry and the functional connectivity between different brain regions could uncover significant differences between the depressed and healthy controls. In a study,[27] EEG features such as the kernel eigen-filter-bank common spatial pattern (KEFB-CSP) and an SVM classifier were combined to perform diagnosis of depression. This method could achieve 80% classification accuracy[26]. Following similar concepts, features such as the power of different EEG frequency bands and EEG alpha interhemispheric asymmetry[28] and the functional connectivity measures such as the synchronization likelihood[29] have been utilized for feature extraction. Later, these features were classified with the logistic regression, naïve Bayesian, and SVM utilized for the classification of the study participants into their respective groups. A summary of all these methods is provided in Table 1.1 and will be discussed in detail in Chapter 2, EEG Fundamentals, in the literature review of related studies.

1.8 ELECTROENCEPHALOGRAPHY-BASED TREATMENT SELECTION FOR DEPRESSION

The EEG/ERP-based methods for prediction of antidepressant treatment efficacy assessment are termed as *EEG/ERP-based predictive biomarkers for MDD*.[21,25] For example, ML-based technique provides an automation during treatment selection for MDD patients, termed "Psychiatrists in the Machine."[56] The technique utilizes an automated computer system that acquires information extracted from a MDD patient's EEG data to prescribe suitable antidepressants. This technique could save time and may help during further standardizing the treatment process for MDD. In a clinic, a professional may use this technique as a second opinion to select suitable medication for a particular MDD patient that also forms the basis of personalized medication.[57]

In the literature, the EEG/ERP-based methods to predict the antidepressant's *treatment selection* have proposed various EEG-based features such as the spectral power computed from the alpha and theta frequency bands. In addition, their mathematical combinations are proposed as an antidepressant treatment response (ATR) index[58] and QEEG theta cordance[59] (details are provided in Chapter 2: EEG Fundamentals). The ATR index-based methods provide results with 70% accuracy for classifying treatment responders (R) and nonresponders.[58] A further study with 25 participants achieved 88% accuracy.[58,60] In the case of QEEG theta cordance, a decrease of prefrontal values of QEEG theta cordance has shown a correlation with the treatment R.[59,61–63] Both the ATR index and the QEEG theta cordance have shown promise. However, these methods have exhibited low specificities (e.g., 70%) which decreases their feasibility for clinical applications. In addition, the related brain mechanisms in support of these measures are less discussed and largely remain unclear.[64]

Referenced EEG (rEEG)[65] involves the selection of suitable treatments based on the evidence provided by a normative database of QEEG patterns. The prescriptions related to certain QEEG patterns are listed. The database has been developed and built over a time frame including various patient's medical records, treatment histories, and related QEEG patterns. The rEEG method allows the MDD patients to go through an EEG recording session during their visit to a clinic. The QEEG patterns are then compared with the recorded patterns from the database. For example, the technique performs comparisons between QEEG patterns of MDD patients and the QEEG patterns saved in the database. The best matched pairs help in selecting a suitable treatment for the MDD patient. The rEEG method has shown better treatment results than the STAR*D study.[65,66] However, its clinical utility has been less studied and may need further research efforts.

EEG has low spatial resolution, therefore techniques such as LORETA (low-resolution brain tomographic analysis) and its variants[67,68] are employed to localize neuronal sources inside the brain. These techniques have contributed toward the development of predictive biomarkers for MDD. For example, LORETA-based studies have reported correlations between rostral anterior cingulate cortex (rACC) activation and the antidepressants' treatment response.[69–72]

Recently, EEG-based ML methods for treatment selection have been proposed to automate the treatment selection process,[56] including for conditions such as schizophrenia and depression.[73,74] In the literature, the ML techniques employed for treatment selection have performed with a maximum 87.5% classification accuracy. However, further replication of these findings with improved classification accuracy (in a larger population) is needed.

ERP-based predictive biomarkers are derived from electrophysiological data captured during the performance of an activity, for example, an oddball task. Previous research studies based on ERP components such as P300 and loudness dependence auditory evoked potentials (LDAEP) have been performed for treatment selection involving MDD patients. For example, the methods have proposed the use of P300 intensities[75] and latencies.[76] Some other studies have utilized auditory evoked potentials known as LDAEP. These studies have reported the associations of the ERP features with treatment efficacy such as the steepness or sharpness in the P300 component and their slopes.[77–79]

In short, the *EEG/ERP-based predictive biomarkers* for MDD have shown less clinical utility due to many factors, especially the EEG findings being seldom replicated. EEG-based neurobiology of MDD has been less explored.[64,80] Moreover, the methods exhibited low performances, specifically, low specificities (less than 70%). Even though EEG has high temporal resolution, the multiresolution decomposition of EEG signals has not been exploited completely for the development of methods for diagnosing MDD. In addition, the QEEG features such as the power of different frequency bands, alpha asymmetry, and synchronization likelihood were integrated. In the literature, the ML techniques employed for treatment selection have performed to a maximum of 87.5% classification accuracy.[73] It is vital to further progress into the development of predictive biomarkers based on electrophysiological datasets. Therefore more solid and systematic research efforts are necessary to develop new techniques.

The research problems that have been identified from the literature and that can be helpful for proposing a framework to assist clinicians to *diagnose* depression and to perform the *treatment selection* for the MDD patients include:

1. The *EEG/ERP-based predictive biomarkers* for MDD have shown less clinical utility that warrants reassessment of current findings with improved accuracy. This will further add to the evidence of the clinical feasibility of EEG/ERP methods for the diagnosis and treatment selection for depression.
2. The EEG-based studies for MDD did not focus on the integration of different EEG/ERP quantities.
3. The EEG-based biomarkers for treatment selection require week 1 data that causes delay in the treatment process.
4. Noise removal and features extraction from the EEG data collected from the MDD patients and healthy controls.
5. Selection of the most noteworthy features that show relevance to the brain regions implicated during depression.
6. Classification of the study participants-based patterns specific to disease conditions.

1.9 DISCUSSION

In this book, a new research method is discussed in detail that provides an improved feature selection and classification method while utilizing pretreatment (week 0) EEG-based features to assist clinicians during diagnosis of depression and selection of antidepressants by focusing on unipolar depressed patients only. The challenge was to develop a method which was invariant to changes in the EEG data due to factors other than the actual neural activity, for example, the artifacts in the EEG data. Therefore the task was subdivided into separate sections starting from EEG data preprocessing, features extraction, feature selection, classification, and validation.

The research results are only applicable for unipolar depressed patients. For example, inclusion of MDD patients with psychotic symptoms is out of scope for this book because of the less-clear neurobiology associated with psychotic depression. Furthermore, in this study, the proposed ML method is validated with 64 study participants including both MDD patients and healthy controls that are specific to the Malaysian population. Hence, this constraint restricts generalization of research results and further warrants clinical trials before being applied in a clinical setting.

In this book, the localization of abnormalities caused by MDD was performed based on investigating the differences of brain activations between the MDD patients and healthy controls involving two methods: (1) the brain source localization technique including the standardized low-resolution electromagnetic tomography (sLORETA) analysis; and (2) the 2D topographic maps of activations. These approaches provide localization of the brain areas that show statistical differences between MDD patients and healthy controls. This signifies an association of the brain areas with the pathology imparted due to MDD. Since EEG has higher temporal resolution and comparatively less spatial resolution than functional magnetic resonance imaging (fMRI); therefore due to the less spatial resolution of EEG data, the EEG-based localization is an estimation of the exact source of the abnormalities.

In this book, an ML scheme, an *Intelligent Treatment Management System (ITMS) for treatment selection (ITMS-TS)*, is presented based on pretreatment EEG data of MDD patients. The ITMS-TS for MDD was developed to predict the treatment outcome efficacy of four different antidepressants—escitalopram (E), fluvoxamine (F), sertraline (S), fluoxetine (Fl)—with similar MOA. The objective was to classify MDD patients as either a treatment respondent or a nonrespondent based on the pretreatment EEG data (because the antidepressants effects on the brain are less clear). The ITMS-TS is restricted to these four types of antidepressants only. Because the inclusion of MDD patients with antidepressants with different MOAs requires an increased sample size, such research is not within the scope of this research work.

This book focuses on the multiresolution decomposition of EEG data based on wavelet transform (WT) analysis. Since the time–frequency decomposition of EEG data can be performed with techniques such as the empirical mode decomposition (EMD) and the short-time Fourier transformation (STFT), hence, to prove the performance of the proposed WT analysis, its performance was compared with the EMD and STFT.

The ITMS for diagnosis (*ITMS-diagnosis*) involved EEG features such as EEG spectral power, EEG alpha interhemispheric asymmetry, and synchronization likelihood. The EEG features were used as input data for the proposed ML scheme, that is, *ITMS-diagnosis*. These features were used as discriminants during the diagnosis process. Moreover, the brain region's connectivity was used as between group differences as diagnostic criterion for the development of the proposed ML framework.

Both qualitative (questionnaires) and quantitative techniques were used to validate the proposed ML framework. For example, in this study clinical assessments were considered as the gold standard involving questionnaires such as the Beck Depression Inventory II (BDI-II) and the Hospital Anxiety and Depression Scale (HADS). The clinical scores (outcomes) were then compared with the predictions generated by the proposed ML framework. The quantitative assessment involved several parametric measures such as classifier accuracies, specificities, and sensitivities. Because the plotting of box-plots require enough statistical data; therefore, 100 times iterations of the 10-fold cross validation were performed.

1.10 SUMMARY

This chapter introduces the basic concepts of depression from the perspective of a layman's point of view such as the different kinds of depression and its subtypes, as well as the challenges in addressing depression as an illness. In addition, the chapter emphasizes the importance of EEG as a modality for automatic identification of MDD patients while explaining the benefits of EEG-based methods for depression diagnosis and treatment efficacy assessment.

References

1. WHO. *Depression: key facts.* <http://www.who.int/news-room/fact-sheets/detail/depression>; 2018.
2. Healthdirect. *Types of depression.* <http://www.mindhealthconnect.org.au/types-of-depression>.

3. WHO. *The Global Burden of Disease: 2004 Update*. Geneva, Switzerland: World Health Organization; 2008:2008.
4. Porjesz, B., & Begleiter, H. (2003). Alcoholism and human electrophysiology. *Alcohol Res Health*, *27*(2), 153–160.
5. Bowden CL. A different depression: clinical distinctions between bipolar and unipolar depression. *J. Affect Disord*. 2005;84:117–125.
6. Association AP. *Diagnostic and Statistical Manual of Mental Disorders (DSM-5®)*. American Psychiatric Pub; 2013.
7. Gartlehner G, Hansen RA, Thieda P, et al. *Comparative Effectiveness of Second-Generation Antidepressants in the Pharmacologic Treatment of Adult Depression. Comparative Effectiveness Review No. 7*. RTI International-University of North Carolina Evidence-Based Practice Center; 2007.
8. Bartels SJ, Dums AR, Oxman TE, et al. Evidence-based practices in geriatric mental health care. *Psychiatr Serv*. 2014;53(11):1419–1431.
9. Simon GE, Khandker RK, Ichikawa L, Operskalski BH. Recovery from depression predicts lower health services costs. *J Clin Psychiatry*. 2006;67:1226–1231.
10. Trivedi MH, Fava M, Wisniewski SR, et al. Evaluation of outcomes with citalopram for depression using measurement-based care in STAR*D: implications for clinical practice. *Am J Psychiatry*. 2006;163:28–40.
11. Acharya UR, Sree SV, Alvin APC, Yanti R, Suri JS. Application of non-linear and wavelet based features for the automated identification of epileptic EEG signals. *Int J Neural Syst*. 2012;22:1–14.
12. Adeli H, Ghosh-Dastidar S, Dadmehr N. A wavelet-chaos methodology for analysis of EEGs and EEG subbands to detect seizure and Epilepsy. *IEEE Trans Biomed Eng*. 2007;54:205–211.
13. Seshadri A, Salim A. Brain Death Evaluation and Determination. In: *Surgical Critical Care Therapy*. Springer, Cham. 2018;61–65.
14. Acharya UR, Faust O, Kannathal N, Chua T, Laxminarayan S. Non-linear analysis of EEG signals at various sleep stages. *Comput Methods Programs Biomed*. 2005;80:37–45.
15. Olofsen E, Sleigh J, Dahan A. Permutation entropy of the electroencephalogram: a measure of anaesthetic drug effect. *Br J Anaesth*. 2008;101:810–821.
16. Acharya UR, Sudarshan VK, Adeli H, Santhosh J, Koh JE, Adeli A. Computer-aided diagnosis of depression using EEG signals. *Eur Neurol*. 2015;73:329–336.
17. Acharya UR, Oh SL, Hagiwara Y, Tan JH, Adeli H, Subha D. Automated EEG-based screening of depression using deep convolutional neural network. *Comput Methods Programs Biomed*. 2018;161:103–113.
18. Leuchter AF, Cook IA, Hunter AM, Korb AS. A new paradigm for the prediction of antidepressant treatment response. *Dialogues Clin Neurosci*. 2009;11:435–446.
19. Iosifescu DV. Electroencephalography-derived biomarkers of antidepressant response. *Harv Rev Psychiatry*. 2011;19:144–154.
20. Huys QJ, Maia TV, Frank MJ. Computational psychiatry as a bridge from neuroscience to clinical applications. *Nat Neurosci*. 2016;19:404–413.
21. Leuchter AF, Cook IA, Hamilton SP, et al. Biomarkers to predict antidepressant response. *Curr Psychiatry Rep*. 2010;12:553–562.
22. Paulus MP, Huys QJ, Maia TV. A roadmap for the development of applied computational psychiatry. *Biol Psychiatry Cogn Neurosci Neuroimaging*. 20161(5):386–392.
23. Adams RA, Huys QJ, Roiser JP. Computational psychiatry: towards a mathematically informed understanding of mental illness. *J Neurol Neurosurg Psychiatry*. 2016;87:53–63.
24. Montague PR, Dolan RJ, Friston KJ, Dayan P. Computational psychiatry. *Trends Cogn Sci*. 2012;16:72–80.
25. Mumtaz W, Malik AS, Yasin MAM, Xia L. Review on EEG and ERP predictive biomarkers for major depressive disorder. *Biomed Signal Process Control*. 2015;22:85–98.

26. Olbrich S, Arns M. EEG biomarkers in major depressive disorder: discriminative power and prediction of treatment response. *Int Rev Psychiatry*. 2013;25:604–618.
27. Liao S-C, Wu C-T, Huang H-C, Cheng W-T, Liu Y-H. Major depression detection from EEG signals using kernel eigen-filter-bank common spatial patterns. *Sensors*. 2017;17:1385.
28. Mumtaz W, Xia L, Ali SSA, Yasin MAM, Hussain M, Malik AS. Electroencephalogram (EEG)-based computer-aided technique to diagnose major depressive disorder (MDD). *Biomed Signal Process Control*. 2017;31:108–115.
29. Mumtaz W, Ali SSA, Yasin MAM, Malik AS. A machine learning framework involving EEG-based functional connectivity to diagnose major depressive disorder (MDD). *Medical Biol Eng Comput*. 2017;1–14.
30. Faust O, Ang PCA, Puthankattil SD, Joseph PK. Depression diagnosis support system based on EEG signal entropies. *J Mech Med Biol*. 2014;14:1450035.
31. Puthankattil SD, Joseph PK. Classification of EEG signals in normal and depression conditions by ANN using RWE and signal entropy. *J Mech Med Biol*. 2012; 12:1240019.
32. Acharya UR, Sudarshan VK, Adeli H, et al. A novel depression diagnosis index using nonlinear features in EEG signals. *Eur Neurol*. 2015;74:79–83.
33. Hosseinifard B, Moradi MH, Rostami R. Classifying depression patients and normal subjects using machine learning techniques and nonlinear features from EEG signal. *Comput Methods Programs Biomed*. 2013;109:339–345.
34. Knott V, Mahoney C, Kennedy S, Evans K. EEG power, frequency, asymmetry and coherence in male depression. *Psychiatry Res Neuroimaging*. 2001;106:123–140.
35. Alhaj H, Wisniewski G, McAllister-Williams RH. The use of the EEG in measuring therapeutic drug action: focus on depression and antidepressants. *J Psychopharmacol*. 2011;25:1175–1191.
36. Zeng LL, Shen H, Liu L, Hu D. Unsupervised classification of major depression using functional connectivity MRI. *Hum Brain Mapp*. 2014;35:1630–1641.
37. Magnin B, Mesrob L, Kinkingnéhun S, et al. Support vector machine-based classification of Alzheimer's disease from whole-brain anatomical MRI. *Neuroradiology*. 2009;51:73–83.
38. Orrù G, Pettersson-Yeo W, Marquand AF, Sartori G, Mechelli A. Using support vector machine to identify imaging biomarkers of neurological and psychiatric disease: a critical review. *Neurosc Biobehav Rev*. 2012;36:1140–1152.
39. Lee J-S, Yang B-H, Lee J-H, Choi J-H, Choi I-G, Kim S-B. Detrended fluctuation analysis of resting EEG in depressed outpatients and healthy controls. *Clin Neurophysiol*. 2007;118:2489–2496.
40. Mohammadi M, Al-Azab F, Raahemi B, et al. Data mining EEG signals in depression for their diagnostic value. *BMC Med Inform Decis Mak*. 2015;15:1.
41. Klöppel S, Abdulkadir A, Jack CR, Koutsouleris N, Mourão-Miranda J, Vemuri P. Diagnostic neuroimaging across diseases. *Neuroimage*. 2012;61:457–463.
42. Erguzel TT, Ozekes S, Tan O, Gultekin S. Feature selection and classification of electroencephalographic signals an artificial neural network and genetic algorithm based approach. *Clin EEG Neurosci*. 2015;46:321–326.
43. McIntyre RS, Cha DS, Soczynska JK, et al. Cognitive deficits and functional outcomes in major depressive disorder: determinants, substrates, and treatment interventions. *Depress Anxiety*. 2013;30:515–527.
44. Jaeger J, Berns S, Uzelac S, Davis-Conway S. Neurocognitive deficits and disability in major depressive disorder. *Psychiatry Res*. 2006;145:39–48.
45. Fingelkurts AA, Fingelkurts AA. Altered structure of dynamic electroencephalogram oscillatory pattern in major depression. *Biol Psychiatry*. 2015;77:1050–1060.

46. Deslandes AC, de Moraes H, Pompeu FA, et al. Electroencephalographic frontal asymmetry and depressive symptoms in the elderly. *Biol Psychol.* 2008;79: 317–322.
47. Stewart JL, Coan JA, Towers DN, Allen JJ. Frontal EEG asymmetry during emotional challenge differentiates individuals with and without lifetime major depressive disorder. *J Affect Disord.* 2011;129:167–174.
48. Cantisani A, Koenig T, Horn H, Müller T, Strik W, Walther S. Psychomotor retardation is linked to frontal alpha asymmetry in major depression. *J Affect Disord.* 2015;188:167–172.
49. Allen JJ, Reznik SJ. Frontal EEG asymmetry as a promising marker of depression vulnerability: summary and methodological considerations. *Curr Opin Psychol.* 2015;4:93–97.
50. Kan D, Lee P, Decrease alpha waves in depression: an electroencephalogram (EEG) study. In: *BioSignal Analysis, Processing and Systems (ICBAPS), 2015 International Conference on.* Kuala Lumpur, Malaysia; 2015:156–161.
51. Saletu B, Anderer P, Saletu-Zyhlarz G. EEG topography and tomography (LORETA) in diagnosis and pharmacotherapy of depression. *Clin EEG Neurosci.* 2010;41:203–210.
52. Coutin-Churchman P, Moreno R. Intracranial current density (LORETA) differences in QEEG frequency bands between depressed and non-depressed alcoholic patients. *Clin Neurophysiol.* 2008;119:948–958.
53. Henriques JB, Davidson RJ. Left frontal hypoactivation in depression. *J Abnorm Psychol.* 1991;100:535.
54. Kemp A, Griffiths K, Felmingham K, et al. Disorder specificity despite comorbidity: resting EEG alpha asymmetry in major depressive disorder and post-traumatic stress disorder. *Biol Psychol.* 2010;85:350–354.
55. Gold C, Fachner J, Erkkila J. Validity and reliability of electroencephalographic frontal alpha asymmetry and frontal midline theta as biomarkers for depression. *Scand J Psychol.* 2013;54:118–126.
56. Moore, S. K. (2011). The psychiatrist in the machine. *IEEE Spectrum,* 48(3) 11–12.
57. Olbrich S, van Dinteren R, Arns M. Personalized medicine: review and perspectives of promising baseline EEG biomarkers in major depressive disorder and attention deficit hyperactivity disorder. *Neuropsychobiology.* 2016;72:229–240.
58. Leuchter AF, Cook IA, Marangell LB, et al. Comparative effectiveness of biomarkers and clinical indicators for predicting outcomes of SSRI treatment in major depressive disorder: results of the BRITE-MD study. *Psychiatry Res.* 2009;169:124–131.
59. Bares M, Brunovsky M, Novak T, et al. The change of prefrontal QEEG theta cordance as a predictor of response to bupropion treatment in patients who had failed to respond to previous antidepressant treatments. *Eur Neuropsychopharmacol.* 2010;20:459–466.
60. Leuchter A, Cook I, Gilmer W. Effectiveness of a quantitative electroencephalographic biomarker for predicting differential response or remission with escitalopram and bupropion in major depressive disorder. *Psychiatry Res.* 2009;169:132–138.
61. Bares M, Brunovsky M, Kopecek M, et al. Early reduction in prefrontal theta QEEG cordance value predicts response to venlafaxine treatment in patients with resistant depressive disorder. *Eur Psychiatry.* 2008;23:350–355.
62. Cook I, Leuchter A. Prefrontal changes and treatment response prediction in depression. *Semin Clin Neuropsychiatry.* 2001;6:113–120.
63. Leuchter AF, Cook IA, Lufkin RB, et al. Cordance: a new method for the assessment of cerebral perfusion and metabolism using quantitative encephalography. *Neuroimage.* 1994;1:208–219.
64. Baskaran A, Milev R, McIntyre R. The neurobiology of the EEG biomarker as a predictor of treatment response in depression. *Neuropharmacology.* 2012;63:507–513.

65. DeBattist C, Kinrys G, Hoffman D, et al. The use of referenced-EEG (rEEG) in assisting medication selection for the treatment of depression. *J Psychiatr Res.* 2011;45:64–75.
66. DeBattista C, Hoffman D, Schiller M, Iosifescu D. Referenced-EEG (rEEG) guidance of medications for treatment resistant depressed patients—a pilot study. In: *Poster no. 228. US Psychiatric and Mental Health Congress.* San Diego, CA; 2008.
67. Pascual-Marqui R, Michel C, Lehmann D. Low resolution electromagnetic tomography: a new method for localizing electrical activity in the brain. *Int J Psychophysiol.* 1994;18:49–65.
68. Jatoi MA, Kamel N, Malik AS, Faye I, Begum T. A survey of methods used for source localization using EEG signals. *Biomed Signal Process Control.* 2014;11:42–52.
69. Pizzagalli D. Frontocingulate dysfunction in depression: toward biomarkers of treatment response. *Neuropsychopharmacology.* 2011;36:183–206.
70. Pizzagalli D, Pascual-Marqui RD, Nitschke JB, et al. Anterior cingulate activity as a predictor of degree of treatment response in major depression: evidence from brain electrical tomography analysis. *Am J Psychiatry.* 2001;158:405–415.
71. Mulert C, Juckel G, Brunnmeier M, et al. Rostral anterior cingulate cortex activity in the theta band predicts response to antidepressive medication. *Clin EEG Neurosci.* 2007;38:78–81.
72. Korb AS, Hunter AM, Cook IA, Leuchter AF. Rostral anterior cingulate cortex theta current density and response to antidepressants and placebo in major depression. *Clin Neurophysiol.* 2009;120:1313–1319.
73. Khodayari-Rostamabad A, Reilly JP, Hasey GM, de Bruin H, MacCrimmon DJ. A machine learning approach using EEG data to predict response to SSRI treatment for major depressive disorder. *Clin Neurophysiol.* 2013;124:1975–1985.
74. Khodayari-Rostamabad A, Hasey GM, MacCrimmon DJ, Reilly JP, de Bruin H. A pilot study to determine whether machine learning methodologies using pre-treatment electroencephalography can predict the symptomatic response to clozapine therapy. *Clin Neurophysiol.* 2010;121:1998–2006.
75. Isıntas M, Ak M, Erdem M, Oz O, Ozgen F. Event-related potentials in major depressive disorder: the relationship between P300 and treatment response. *Turk Psikiyatri Derg.* 2012;23:33–39.
76. Kalayam B, Alexopoulos GS. Prefrontal dysfunction and treatment response in geriatric depression. *Arch Gen Psychiatry.* 2011;56:713–718.
77. Gallinat J, Bottlender R, Juckel G, et al. The loudness dependency of the auditory evoked N1/P2-component as a predictor of the acute SSRI response in depression. *Psychopharmacology (Berl).* 2000;148:139–143.
78. Linka T, Müller B, Bender S, Sartory G, Gastpar M. The intensity dependence of auditory evoked ERP components predicts responsiveness to reboxetine treatment in major depression. *Pharmacopsychiatry.* 2005;38:139–143.
79. Linka T, Sartory G, Wiltfang J, Müller B. Treatment effects of serotonergic and noradrenergic antidepressants on the intensity dependence of auditory ERP components in major depression. *Neurosci Lett.* 2009;463:26–30.
80. Stephan KE, Mathys C. Computational approaches to psychiatry. *Curr Opin Neurobiol.* 2014;25:85–92.

CHAPTER

2

Electroencephalography Fundamentals

2.1 WHAT IS ELECTROENCEPHALOGRAPHY?

As shown in Fig. 2.1, the bulk of neuronal activity from various on-scalp locations is captured as an ensemble of electrical records, termed as electroencephalography (EEG) data. In most cases, EEG is a noninvasive and safe modality for capturing the human brain's neural activity with high temporal resolution [millisecond (ms)].[1] Therefore, EEG has made the online monitoring of epileptic patients and the prediction of occurrence of an epileptic seizure activity a reality.

Resting-state EEG data involve recording during two physiological conditions: eyes closed (EC) and eyes open (EO). More importantly, resting-state EEG data are useful for studying the behavior of default brain networks and their associated background activities. On the other hand, event-related potential (ERP) data involve brain electrical activities recorded during experimental activity time locked to an event. The ERP components, such as P300, can be extracted from the ERP data and are considered as indices of cognition, attention, and risk markers for depression.[2]

2.2 ELECTROENCEPHALOGRAPHY APPLICATIONS

Usually, brain electrical activity is measured in microvolts and EEG data are direct representations of human brain activity. Many EEG patterns are expected to appear during certain physiological states, for example, EEG patterns during sleep have shown specific patterns as mentioned in Fig. 2.2. As seen in Fig. 2.2, brain death is represented as a straight line which means that no electrical activity is detected.

EEG-Based Experiment Design for Major Depressive Disorder
DOI: https://doi.org/10.1016/B978-0-12-817420-3.00002-3

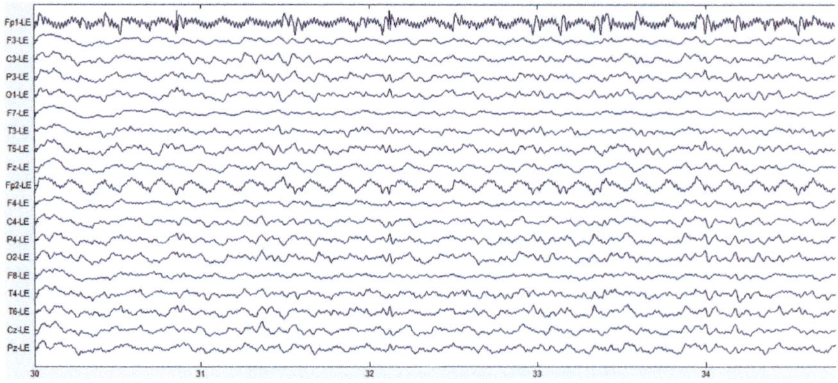

FIGURE 2.1 A small trace of EEG data.

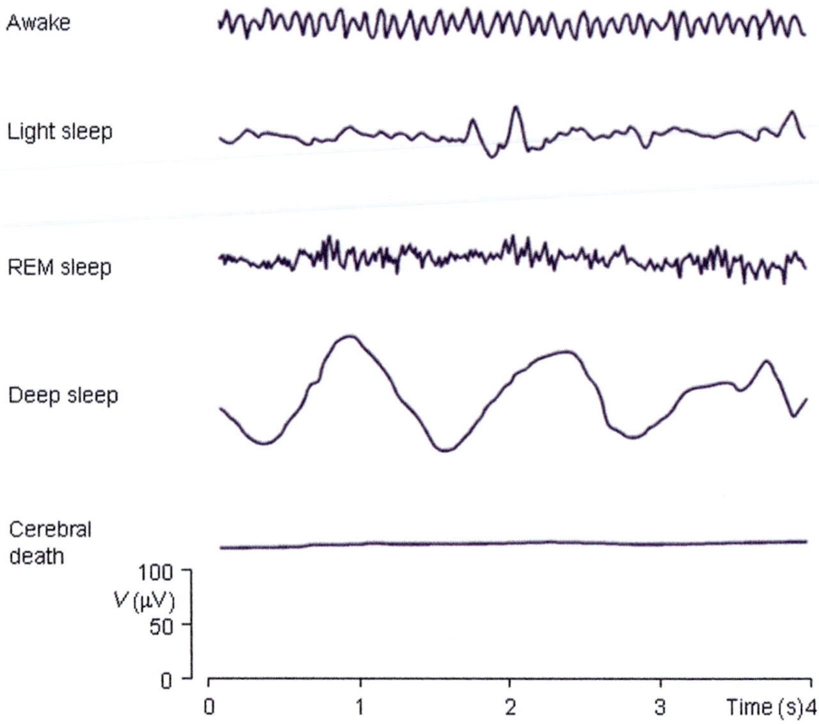

FIGURE 2.2 EEG patterns specific to various physiological conditions.

Other applications include:

1. monitoring alertness, coma, and brain death;
2. locating areas of damage following head injury, stroke, tumor, etc.;
3. testing afferent pathways (by evoked potentials);
4. monitoring cognitive engagement (alpha rhythm);
5. producing biofeedback situations, alpha, etc.;
6. controlling anesthesia depth ("servo anesthesia");
7. investigating epilepsy and locating seizure origin;
8. testing epilepsy drug effects;
9. assisting in experimental cortical excision of epileptic focus;
10. monitoring human and animal brain development;
11. testing drugs for convulsive effects; and
12. investigating sleep disorders and physiology.

2.3 ELECTROENCEPHALOGRAPHY FREQUENCY BANDS

Moreover, EEG data are composite in nature and can be decomposed into different frequency bands, such as: Delta (0.5–4 Hz); Theta (5–8 Hz); Alpha (8–12 Hz); low Beta (12–18 Hz); high Beta (19–30 Hz); and Gamma (30–70 Hz).

2.3.1 Delta Band

The delta band (0.5–4 Hz) involves diffused slow waves and is generally seen during sleep, injury, coma, and brain trauma. In addition, delta waves are nonsinusoidal and wandering and are often characterized as nonrhythmic activity. The underlying mechanisms could be due to slow neuronal mechanisms. It is often seen after epileptic seizures. Sources other than the brain, such as skin potentials and slow cortical potentials, can show up as delta band.

2.3.2 Theta Band

Like the delta band, the theta band (5–8 Hz) consists of nonsinusoidal waves, such as square waves, and is often mediated by subthalamic connections to the cortex. Theta waves often indicate either distractibility or a deep inward awareness or lack of presence. Theta waves are common in attention deficit disorder and other types of disorders where brain excitability is not well regulated. Theta waves are often associated with memory retrieval and extreme creativity. Moreover, experienced meditators tend to exhibit theta waves during meditation, which indicates an internalized state and a state of inward focus. Furthermore, during epilepsy and attention deficit disorder excessive theta waves are seen.

2.3.3 Alpha Band

The alpha band (8–12 Hz) is the resting rhythm of the visual system and is found to be maximal at the occipital area. Alpha waves are highly sinusoidal and smooth indicating that the visual system is at rest and is not processing information. It could explain other mechanisms, such as reverberation between the thalamus and cortex as well as memory scanning. The alpha band does occur in other brain locations such as the temporal lobe. Moreover, connectivity measures are important with alpha waves, such as coherence, synchrony, and asymmetry. These measures of alpha are of tremendous diagnostic importance. For example, alpha synchrony may be used as a biofeedback parameter even in people who are perfectly normal and wanted to improve their mental performance. On the contrary, there are variants of high alpha that are abnormal. For example, if you see large amounts of diffuse alpha in an individual, and that alpha is seen to drone on and on, this would indicate an anxiety disorder.

2.3.4 Low Beta Band

The low beta band (12–15 Hz) is often termed as 14 Hz, as mentioned by Michael Tansey who is a pioneer in this field. He made use of a filter with a pass range of 13–15 Hz in order to center on 14 Hz. Other researchers have used 12–16 Hz. This rhythm, when observed over a motor strip (C3, C4, and Cz), is called sensory motor rhythm (SMR). SMR is the alpha waves of the motor system. It is maximal at C3, C4, and Cz, and when the body is still. Dr. Barry Sterman discovered this rhythm in cats who were sitting still. SMR is maximum when the brain intends to remain still. Hence, the SMR can be used as a parameter during biofeedback therapy. SMR deals with intentions, that is, that which an individual intends or plans to do. When it is uptrained, it was found that individuals become quiet, still, relaxed, are focused and remain in a concentrated state. The state in which lots of SMR is being produced is studious, quiet, healthy, and pensive state. One can solve math problems, read, or sit through a lecture in that state (i.e., your legs are not jiggling, you are not shuffling around in your chair, you are not wandering about during lunch. or when class is over).

The thalamic–cortical reverberations between the motor strip and thalamus take about 80 ms in one direction and then 80 ms to get back. The distance is much shorter than the distance from the thalamus to the back of head (occipital). Therefore, SMR is slightly faster than alpha because of the fact that it is simply a shorter trip to go from the thalamus to the central lobe compared to going from the thalamus to the occipital lobe.

2.3.5 Beta Band

The beta band (15–20 Hz) occurs during a thinking state. As we move to higher frequencies, waves become more and more localized. For example, alpha could involve both hemispheres and one could have synchronous alpha for both hemispheres. Low beta is typically more localized to one hemisphere or the other (low beta synchrony across hemispheres is not done. In fact, it is asymmetrical between hemispheres.) Beta is even more localized in a certain area. The reason is that the connections in the brain that produce beta waves are more cortical–cortical. It's one part of the cortex talking to a neighboring part in the cortex. Hence, beta waves tend to be localized where work is being done. Beta activity may be associated with memory recall when reading (increased activity over the motor strip). If a normative amount of beta does not occur it means that there is underactivity (individual may not be sharp enough, feeling depressed, etc.). In other words, beta is a sign of activation.

2.3.6 High Beta Band

The high beta band is usually from 20 to 30 Hz. In some case, it could go up to 35 Hz. At high beta, muscular activity can sneak in electromyography (EMG). The only way to tell the difference between high beta and muscle activity is by looking at the waves. Muscle activity is more like droning activity while high beta is more like waxing and waning, that is, true sinusoidal. High beta is associated with intensive thinking, such as worrying about things like solving a hard math problem. Generally, guard (stop) bands are used here so that artifacts, like teeth grinding, muscular activity, etc., do not leak down to alpha. With kids, it is extremely important to observe them. They might make slight faces when nervous or for other reasons and hence produce the desired waves and get rewarded. Therefore, it is important to explain to them in detail and then also to observe them, otherwise, false data may be recorded or observed. One needs to tell them to be quiet, still, relaxed and to listen and look, and only then to ring the bell (for example).

2.3.7 Gamma Band

The gamma band is typically from 40 Hz and above. Researchers have gone up to 64 Hz and still referred to it as gamma. The gamma band indicates high levels of sensory and perceptual binding and may be an indicator that binding has occurred (researchers are unsure). Binding can be explained using a simple example; let's take a pen—one can move it in their fingers (motor interaction), see it from any side, hide behind, or just feel it—but they know it is a single object: it is a pen. There is no confusion in the brain as to what it is, and the brain sees it as a single entity. On the contrary, a small child does not have this ability. When an object has gone out of sight, it is gone. When it reappears, it is something new to them. An adult can smell the ink of a pen and can hear if it has a button, etc. This means that there is information coming from all the sensory mechanisms (sight, sound, smell, touch, taste, memory), which informs a person about the pen. This indicates the binding among the sensory, perceptual, and memory systems inside the human brain.

2.4 ELECTROENCEPHALOGRAPHY RECORDING TECHNIQUES

EEG recording has fundamental importance for recording quality data because data quality has direct effects on analysis and subsequent inferences. The EEG recording system consists of electrodes (EEG sensors) with conductive media and amplifiers with suitable filter settings.

2.4.1 Electroencephalography Sensor's Location

As shown in Fig. 2.3, the location of EEG sensors over the scalp follow a standard electrode placement known as the international 10–10 and 10–20 systems.[3] The sensors are labeled according to their spatial locations, for example, the frontal electrodes might have labels such as Fp1, Fp2, Fpz, F3, F4, F7, and F8; electrodes in the parietal are labeled as P3, P4, and Pz; the occipital electrodes are labeled as O1, O2, and Oz, etc. Most vendors have provided predesigned caps with operational manuals. The end users or researchers can achieve correct placements of the sensors by correctly observing the operational manual.

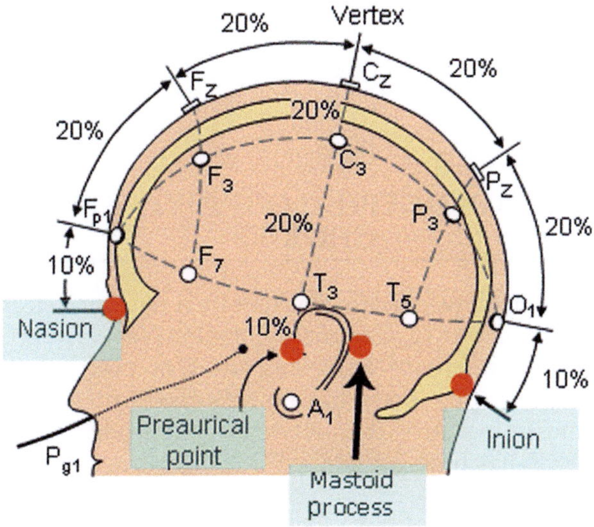

FIGURE 2.3 19 scalp EEG sensor locations.

2.4.2 Electroencephalography Sensors and Conductive Media

EEG recording is impossible without establishing a contact between the sensors and the scalp surface, this is achieved using a conducting media. A connection can be achieved through a conducting material such as a gel or water (in the case of a hydrosensor). The use of gel or water depends on the application type. For example, the gel-based electrodes are preferred over the hydrosensor-based electrodes during sleep studies. Because the gel-based electrodes can work efficiently for longer hours of recording than the hydro-sensor-based electrodes. On the contrary, the hydrosensor-based electrodes have shorter setup time than the gel-based electrodes and preferred for small duration experiments such as for 1 to 2 hours.

2.4.3 Electroencephalography Amplification

The use of amplifiers for amplifying a weak EEG signal is quite common during EEG recording. An amplifier not only provides amplification but also performs analog-to-digital (A/D) conversion as well. Usually, an amplifier has these settings: low and high frequency range, sampling rate and number of channels, type of channels, etc. For example, the low and high frequencies could be 0.5–70 Hz (covering delta, theta, alpha, beta, and gamma bands), while the sampling rate can be

128, 256 Hz, etc. The number of channels can vary from as low as 1 channel to 1024 channels depending on the application. For example, if source localization needs to be performed, the lowest number recommended is 64 channels.

2.5 ELECTROENCEPHALOGRAPHY REFERENCE CHOICES

EEG data are electrical signals that require a reference electrode to be placed on the human body. Commonly, the link-ear reference, Cz reference, and bipolar reference are being used in practice. It has been assumed that the human body has no reference with a potential zero. Therefore, most recently the reference electrode standardization technique (REST) has been introduced that has been considered to be at potential zero, theoretically.

Recently, the choice of an appropriate EEG reference has been highlighted as a critical issue, especially during EEG-based clinical decision making, such as the automatic diagnosis of epilepsy, depression, and schizophrenia. In the literature, studies based on EEG reference choices have reported differences in EEG-derived measures due to distinct EEG references, for example, bias in the computed head surface integral.[4,5] In addition, differences in the computation of the frequency spectrum based on EEG datasets are reported.[6] Moreover, one of the initial studies that investigated the influence of the reference on the experimental effects between two groups[7] and EEG-based connectivity analysis[8] have shown similar effects. Furthermore, the effects of EEG reference choices on information–theoretic measures of the complexity are reported.[9] In a recent study, EEG-based brain network analysis shows discrepancies just because of EEG reference choices.[10] To further elaborate this point, a comparative study of average, linked mastoid, and REST references for ERP components acquired during functional magnetic resonance imaging (fMRI) has been performed.[11] The authors have recommended REST for EEG/ERP analyses. Similarly, REST has been recommended for the analysis of ERP component P300.[12] In a different study, the authors recommended REST to be the first choice of re-referencing, while average reference (AR) may be an alternative option in the case of high-level sensor noise.[13] In short, all these evidences provided a clear picture of the influence of reference choices on EEG data analysis.

For the sake of explanation, three different references, linked-ear (LE), AR, and REST, have been utilized. A brief summary of each reference is provided in Sections 2.5.1–2.5.3.

2.5.1 Linked-Ear Reference

According to the LE reference, as shown in the mathematical formulation provided in Eq. (2.1):

$$\begin{aligned} V_{LE} &= V - T(M_{left} + M_{right})/2 \\ V_{LE} &= (V - TM_{left}) - T(M_{right} - M_{left})/2 \\ V_{LE} &= V_{CM} - TV_{CM-right}/2 \end{aligned} \tag{2.1}$$

where the data matrix V and V_{CM} with size $n \times k$ represents the scalp potential recordings with n electrodes and k samples. T is a column vector with size $n \times 1$ and all its entry values are set at unity. The $V_{CM\text{-right}}$ with size $1 \times k$ is the potential recorded as right mastoid with the left mastoid as the reference, that is, the recording in the V_{CM} corresponding to the right mastoid electrode. M_{left} and M_{right}, with size $1 \times k$, represent the scalp potential of the left and right mastoids when referenced to a neutral point, respectively.

2.5.2 Average Reference

EEG data recorded with AR records potential according to the average of voltage potentials at all scalp sensors at each time point. Mathematically, it can be expressed as (Eq. 2.2):

$$\begin{aligned} V_{AR} &= V - T_{Mean}(V) \\ &= V_{CM} - T_{Mean}(V_{CM}) \end{aligned} \tag{2.2}$$

where the data matrix V and V_{CM} with size $n \times k$ represents the scalp potential recordings at n electrodes and k samples. T is a column vector with size $n \times 1$ and all its entry values are set at unity. Where mean (*) denotes the spatial average over all recording channels at each temporal sample point.

Based on the provided formulae, the recordings with the other references, such as the LE and AR, are readily obtained from the actual recordings V_{CM}. The formulae can be implemented with Matlab or any other programing software.

2.5.3 Reference Electrode Standardization Technique

The REST reference has been considered to be at potential zero, theoretically. Mathematically, it can be computed as (Eq. 2.3):

$$V = GS \tag{2.3}$$

while for the contralateral mastoid reference (CM) recording V_{CM}, we have (Eq. 2.4):

$$V_{CM} = G_{CM}S \tag{2.4}$$

where the G and G_{CM} are the transfer matrices determined by the head model, source configuration, electrode montage and reference, infinity, and left mastoid, respectively. S represents actual sources. Based on the previous equations, we can write (Eq. 2.5):

$$V = GS = G\left((G_{CM})^{+}V_{CM}\right) = RV_{CM} \tag{2.5}$$

where $(*)^{+}$ is the Moore–Penroose generalized inverse, and R is the transfer matrix. Based on these equations, there is no need to know the actual source S; however, the transfer matrices G and G_{CM} can be defined as being the transfer matrix from the equivalent distributed source of the actual sources. Thus, the G and G_{CM} are determined by the head model.

2.6 ELECTROENCEPHALOGRAPHY ARTIFACTS

EEG artifacts can be of diverse types, such as electro-ocular (EOG) (eye blinks, and horizontal and vertical eye movements), electromyography (EMG) (muscular activities), electrocardiogram (ECG) (heart activities), gait-related movement artifacts, power line noise (50 or 60 Hz signal component), and noise, due to floating electrodes, such as a loose connection with the scalp or high impedance electrodes, discontinuity in the data, or artifacts due to electric devices interference. Some of these artifacts are more prominent and can cause severe damage to the EEG data, and are described in Sections 2.6.1–2.6.4.

2.6.1 Ocular Artifacts

Fig. 2.4 shows an ocular artifact (OA) in EEG data. Ocular artifacts are considered as the most important artifacts during EEG artifact reduction because they occur frequently and the amplitude of EOG artifacts is even higher than that of EEG data. Ocular artifacts involve both the eye-blink and the eye movements (horizontal and vertical), and normally occur in low frequency bands. Sources of ocular artifacts include the existence of a dipole model. For example, the eye forms an electric dipole where the cornea is positive and the retina is negative. When eye movement occurs the electric field around the eye changes producing an electric signal, this is known as an ocular artifact. Many EEG artifact reduction methods have been proposed to remove EEG

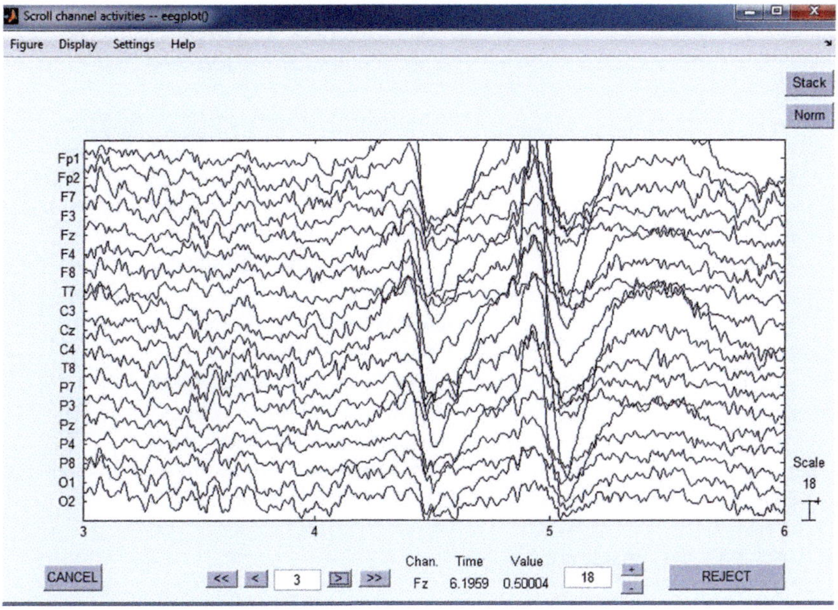

FIGURE 2.4 EEG ocular artifact (OA).

artifacts. For example, Gasser and coworkers found that eye movements increased the power of low frequency bands in schizophrenic patients, and that they also decreased the chance of finding significant differences between the schizophrenic and control groups.[14]

2.6.2 Muscle Artifacts

Fig. 2.5 shows muscle artifacts that often occur during EEG recordings and that have been characterized as high frequency bursts. Artifact sources could be head movement, mouth clenching, swallowing of saliva, etc. Furthermore, the topographic distribution of fast EEG activity associated with muscle contamination showed maximal values over frontotemporal areas and was near zero in the remaining cortical regions. EEG segments contaminated with muscle activity suffered a drastic enhancement of spectral power above 50 Hz. Many methods have been developed for muscle artifact reduction.[15,16]

2.6.3 Line Noise Artifacts

Fig. 2.6 shows a power spectrum of an EEG line noise artifact peaked at 50 Hz, which could confound an EEG signal. The possible sources of

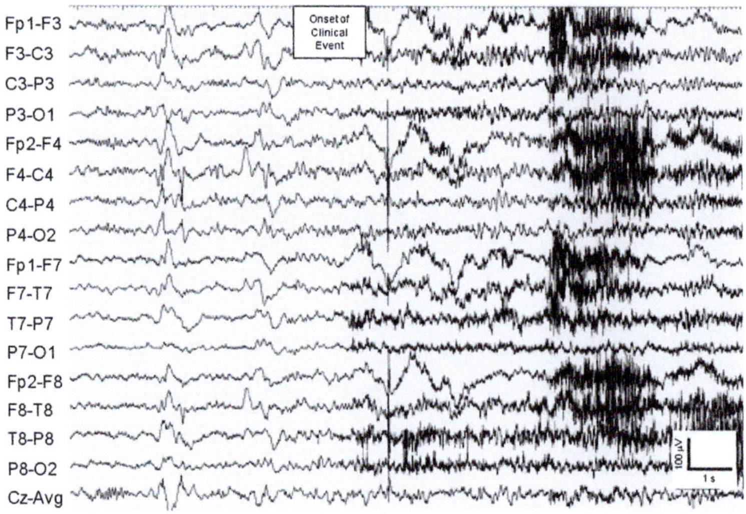

FIGURE 2.5 EEG muscle artefact.

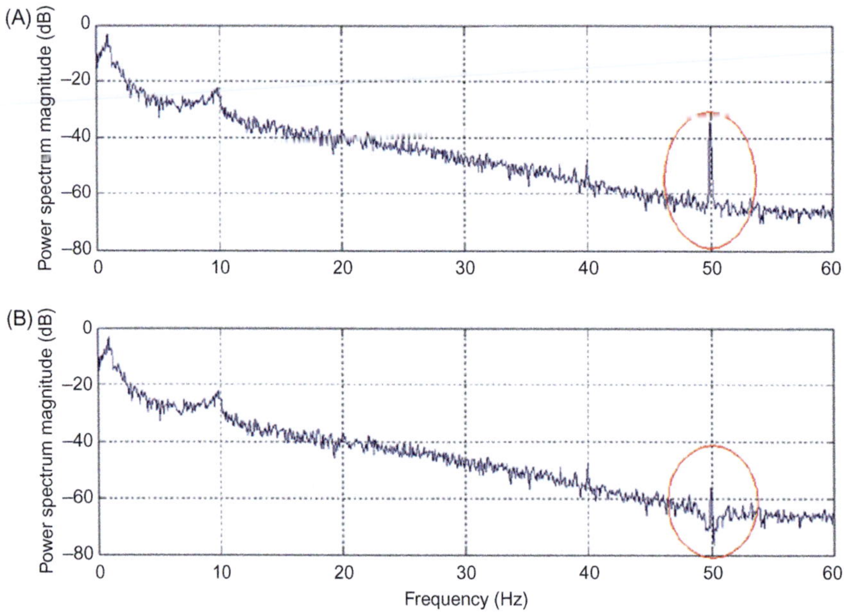

FIGURE 2.6 EEG line noise artefact: (A) power spectrum magnitude of EEG with artifact (EEG + EOG + ECG + line frequency) and (B) power spectrum magnitude of EEG without line frequency.

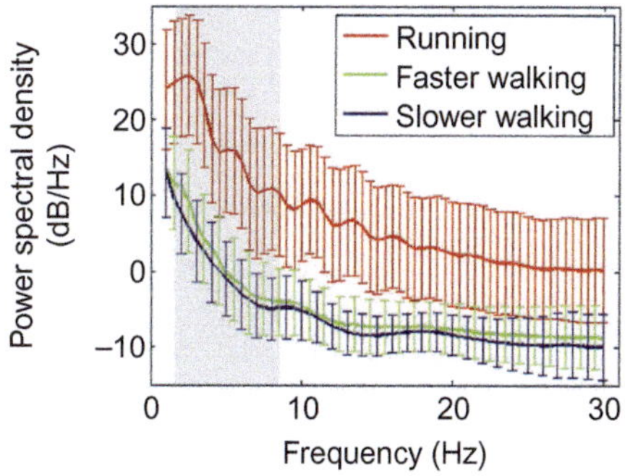

FIGURE 2.7 EEG gait-related artefact.

a line noise artifact include the noise produced from the electric power supply to the system. Commonly, a notch filter designed for 50 or 60 Hz can remove the EEG line noise artifact, successfully.

2.6.4 Gait-Related Motion Artifacts

Fig. 2.7 shows a power spectral view of an EEG signal confounded with gait-related artifacts. Possible sources are running, or fast or slow walking. The power spectrum shows the amplitude of artifacts during running, fast walking, and slow walking. The difference is quite significant in terms of amplitude. Recently, many studies have been performed to correct data confounded with gait-related artifacts.[17]

2.7 ELECTROENCEPHALOGRAPHY-BASED METHOD FOR ARTIFACT REDUCTION (ELECTROENCEPHALOGRAPHY PREPROCESSING)

2.7.1 Analog Methods

Analog methods mainly involve the use of an analog circuit, such as a potentiometer, to subtract subject specific EOG activity from EEG data during recording. For example, one of the initial papers verified that involuntary eye movements during the click-tone interval consistently generated potential shifts, which spread from the corneo-retinal dipole to the scalp electrodes, thereby, contaminating the CNV by a factor of 23%.[18] Another study concluded that the blink and vertical eye movement artifact fields are quantitatively different in terms of their transmission to the scalp.[19]

As a solution,[20] described one of the first online methods for reducing OA in CNV recordings using a simple potentiometer arrangement.

TABLE 2.1 EEG Artifact Suppression Based on Simple Subtraction Technique

Study	Artifact Type	Method
20	Ocular artifact	Analog method involving a potentiometer
21	Eye movement	Circuit-based analog method
22	Eye movement	A simple circuit that effectively removes artifacts
23	Eye movement and ECG	Analog method involving a potentiometer

The vertex EEG signal was referred to as the center terminal of the potentiometer. The other terminals of the potentiometer were connected to a frontal EEG electrode and linked mastoids. The corrector was calibrated, prior to recording, by adjusting the center terminal while the subject moved his eyes repetitively until there was no trace of OA in the LEG. The device was then left at this setting during recording.

An earlier paper has suggested removing different fractions of artifact ratios from EEG data because of the different effects on the EEG data.[19] This was later confirmed by Picton (Picton 2000). In this context, the first attempt was made by Girton and Kamiya.[21] The method was based on subtracting a fraction of both the vertical and horizontal EOG signals from the EEG signal to be corrected. Similar work has been done by other researchers as well.[22] The process seems clear; however, the absence of a true reference may effect the process of artifact correction.

In the study,[23] artifact signals are measured with a special set of electrodes and are combined in order to subtract the artifacts from contaminated EEG records. The subtraction ratios are determined as those that minimize the variance in the resulting corrected EEG signal. Table 2.1 shows a summary of methods.

2.7.2 Linear Filtering Methods

Linear filtering methods mainly employ the manipulation of frequencies in the composite EEG signal. For example, filtering EEG data allows for the analysis of brain activity in different frequency bands.[24] EOG artifacts usually have a low frequency profile. Applying a suitable filter involving the suppression of small frequency components could potentially remove EOG from EEG data. For example, in his paper,[25] Gotman et al. concluded that total elimination of activity above 25 Hz would eliminate most of the EMG activity, with a minimal risk of eliminating rhythmic cerebral activity. In this context, a 4-pole analog filter was designed to remove EMG artifacts.[26] Furthermore, an improvement was achieved by designing a 6-pole digital filter.[27] The authors claim that

TABLE 2.2 EEG Artifact Suppression Based on Linear Filtering Technique

Study	Artifact Type	Method
25	EMG	Digital filtering
26	EMG	Analog filtering
27	EMG	Digital filtering

the advantage of a 6-pole switched-capacitor filter is that it does not require a complex design with high precision RC components.

However, due to the overlapping of artifact and neuronal data, linear filtering could also remove the actual EEG data representing the underlying brain activity. Analog filters are less expensive but more difficult to design than digital filters. Table 2.2 provides a summary of linear filtering methods used for the removal of EMG artifacts from EEG data.

2.7.3 Regression-Based Methods

Regression-based methods provide some major improvements on traditional analog methods. Hence, these methods are quite famous for removing ocular artifacts from EEG data. Initially proposed methods involve both time-domain methods[28] and frequency-domain methods.[29] More specifically, the regression function calculates "B," the proportion of one variable explained by another variable. For example, the effect of EOG on EEG can be explained by "B." In the study,[30] a further validation of regression methods was performed on ERP data with the conclusion that regression provides improved results compared to conventional rejection of artifact data.

Despite the applications of regression-based methods, there are many reservations, like the fact that most regression methods are developed for EOG artifacts. In other words, the regression-based methods are less accurate for muscular artifacts than for the EOGs. In addition, regression-based methods require a separate reference signal for the activity that needs to be corrected from raw EEG data and can be utilized as a regressor because the muscle artifacts cannot easily be used as reference signals. In some cases, the calculation of B has been added with random error. Some solutions have been proposed, such as calculating more than one B and either performing mean, median, or trimming methods, where the lowest and highest 10% of B's are ignored.[30]

In addition, regression procedures consider that EOG signals are not correlated with EEG signals.[31] This assumption is generally wrong. A possible remedy for such a problem is the use of low-pass filtering before applying a time-domain regression.[14] In a later study, it was concluded that ocular artifacts affect the alpha and beta bands as well.[32] In this context, the application of a low-pass filter will not be a complete solution. Moreover, regression methods are also confounded due to bidirectional contamination. The idea of bidirectional contamination is that EOG and EEG can affect each other, and hence, both should be considered during artifact reduction with regression techniques. Table 2.3 summarizes various methods involving regression.

TABLE 2.3 EEG Artifact Suppression Based on Regression Methods

Study	Artifact Type	Comments
28	Eye movement	Quilter used the "least squares" regression function. The regression function calculates B.
29	Eye movement	To reduce "correction phase error," the filtering out of EOG frequencies above 6 Hz has been suggested.
30	EOG	Trimming of B has been proposed, such as discarding the lowest and highest 10% of B.
33	EOG	Gratton mainly proposed that the blink and saccade voltages propagate differently.
34	EOG correction in frequency domain	Transfer of eye activity to EEG may have frequency dependent amplitude; therefore the regression formula is used in frequency domain.
35	EOG	The propagation of ocular potentials across the scalp on a biophysical basis.
36	EOG	Comparison between time and frequency domain regression methods.
37	EOG	In event-related potential, to reduce coherence between eye-blink activity and ongoing EEG, VEOG and EEG are averaged on eye blinks. This yields a high reliability and validity of regression factors, determined per day, subject, and lead.
38	Muscle artifact	The regression method is extended for correction of muscle artifact correction.
39	EOG	Simple and multiple lag regression in time domain and frequency domain have been compared and found equally well in reconstructing EEG.
14	EOG	Gasser emphasis the use of correction methods instead of manually selecting the artifact-free epochs.
40	EOG	Aligned-artifact averaging (AAA) solution.
41	EOG	The validation of the regression method is based on blind scoring, which is carried out by an expert EEG analyst.

2.7.4 Dipole Modeling-Based Methods

Dipole modeling-based approaches involve dipoles with different locations and orientations for EOG and EMG activities.[42,43,44] The idea is to model either the artifact topography or the brain topography and perform the subtraction of the artifact from the recorded data in order to achieve artifact-free data. It involves a head model. Normally, artifact topography is modeled because it is easier to estimate than brain activity. However, dipole-based methods are normally confounded by distortion due to the head model. Berg and Scherg in their paper have claimed to eliminate this effect.[42]

The multiple source eye correction (MSEC) method is a variant of dipole-based methods.[45] The method involves modeling eye artifacts that can be further used to model and correct brain activity. In general, eye artifacts can be modeled with a simple regression method. Brain activity is modeled in the presence of eye artifact topography and the choice of a head model. As the MSEC method may involve PCA for the generation of artifact topographies; therefore, the artifacted EEG data needs to be free from linear trends and artifacts due to loose connection problems. This is because the PCA may not perform efficiently against these kinds of artifacts and could confound the artifact correction process. The MSEC approach can be implemented in different versions such as surrogate MSEC. Surrogate MSEC involves the placement of dipoles at strategic locations inside the brain. A through discussion on MSEC method and its variants is out of scope here and can be found else where[45].

The preselection and the spatially constrained ICA (SCICA) methods were proposed by Ille in his papers[46,47] and have shown improved versions that may involve the ICA and PCA techniques. Comparable to MSEC, artifact topographies are derived from single or averaged artifact prototypes. To model brain signal topographies, two novel concepts are introduced: the preselection approach and the spatially constrained ICA (SCICA) method. In the preselection approach, a relevant number of eigenvectors is extracted from an artifact-free subset of the data segment. The subset is obtained by excluding sample vectors from the original data segment that exceed a certain amplitude or correlation with the predefined artifact subspace. In SCICA, brain signal topographies are estimated from the whole data segment. SCICA uses prior knowledge about the artifact topographies and combines this spatial information with the temporal-statistical strategy of ICA to estimate brain signal topographies (see Table 2.4).

2.7.5 Blind Source Separation

Blind source separation (BSS) techniques have been widely utilized to suppress ocular artifacts and are based on the linear decomposition

TABLE 2.4 EEG Artifact Suppression Involving Dipole Model-Based Methods

Study	Artifact Type	Method
42	EOG	Spatio-temporal dipole modeling
43	EOG	Source dipoles and source components
44	EOG	Source dipoles and source components
45	EOG	Multiple source eye correction (MSEC) method
46	EOG	Preselection methods and SCICA
47	EOG	Preselection methods and SCICA

of EEG and EOG recordings into source components. These component-based methods permit the identification of the sources that cause artifacts, and then the reconstruction of EEG recordings without these sources implicating clean EEG data. In this chapter, different EEG-based artifact reduction methods are discussed, including principal component analysis (PCA), independent component analysis (ICA), and canonical correlation analysis (CCA).

While comparing different BSS techniques for all types of contamination, PCA is a strong performer when contamination is greater in amplitude than the brain signals, whereas other algorithms, such as second-order blind inference and infomax are generally better for specific types of contamination of lower amplitude.[48]

2.7.6 Principal Component Analysis

Lins et al. proposed the use of PCA to remove EOG artifacts.[44] More specifically, PCA transforms a multivariable data set into components that are spatially orthogonal. However, the main drawback of PCA is the assumption of orthogonality between neural activity and ocular artifacts and this is not generally true. For example, PCA-based removal of EOG activity from EEG is difficult, especially when both have a similar magnitude.[49] The authors developed a variation of original PCA in which the factors that reconstruct the modified EEG from the original are stored as a matrix. This matrix is applied to multichannel EEG at successive times to create a new EEG continuously in real time, without redoing the time-consuming SVD. This matrix acts as a spatial filter with useful properties.

PCA assumes that different sources in EEG data are uncorrelated; however, the condition of independence cannot be met by PCA. Studies[50] compared PCA with ICA and proved that PCA is better than ICA. Casarotto utilized PCA to reduce ocular artefacts in single trial ERPs recorded in normal and in dyslexic children.[51] Table 2.5 provides a summary of the method involving PCA.

TABLE 2.5 EEG Artifact Suppression Involving PCA

Study	Artifact Type	Method
44	EOG	PCA
49	Ocular movement electrocardiographic artifacts	A variation of original PCA methods which only employ a matrix that acts as a spatial filter.
50	Event-related potential components during correct and incorrect responses	PCA
51	Ocular artifacts	PCA
52	ECG	sPCA

2.7.7 Independent Component Analysis

As introduced by Makeig,[53] ICA is considered as the most effective component-based procedure for removing artifacts from raw EEG data. Regression methods required a clean EOG reference signal to clean artifact contaminated data. On the contrary, the ICA method is based on a spatial filtering concept that does not need a reference signal.

Many sophisticated techniques have been developed for implementation, such as JADI, INFOMAX,[54] etc. In contrast to PCA, ICA assumes that the component sources must be uncorrelated and statistically independent. These assumptions seem more rational because of the independent nature of ocular and neural sources. Initially, manual selection of artifact contaminated independent components (IC) is challenging and requires knowledge or expertise and renders the procedure subjective. To overcome this subjectivity, studies have been carried out using a completely automatic artifact correction process based on statistical properties, like kurtosis, entropy, or correlation with EOG channels.[55,56] Furthermore, opposite results were obtained: in the former study, the best performance was obtained with an ICA algorithm,[56] while in the latter, ocular artifact correction methods based on regression and PCA performed better than ICA-based correction procedures.[57] Moreover, another study defended regression-based methods against BSS when the number of EEG leads was small.[41] In a different study, a reduced number of electrode configuration or even without EOG leads (where a linear regression approach and automatic procedures based on correlation with EOG channels are not applicable) have been analyzed.[58]

However, automatic selection of the ICA component has proven challenging, which renders the process impractical for large datasets. More recently, objective criteria were developed to solve this issue. In addition, ICA cannot decompose linear trends or data recorded with lose electrodes. Hence, before applying ICA on raw data, the data should be clean from linear trends or artifacts due to lose sensor connections. Table 2.6 has summarized different studies that have utilized ICA as the main method for artifact correction.

As previously mentioned, a serious drawback of the ICA method is the manual selection of artifact-related ICA components because the selection requires knowledge and experience of different artifact types for their correction. Unfortunately, subjectivity cannot be avoided with the manual process. Hence, the use of objective criteria for the selection

TABLE 2.6 EEG Artifact Suppression Involving ICA

Study	Artifact Type	Method
53	EOG	ICA for continuous EEG data
59	EOG	The ICA algorithm has been extended for event-related potential (ERP) data.
60	EOG	Separation of ICs based on kurtosis of their amplitude distribution over time, thereby distinguishing between strictly periodical signals, regularly occurring signals, and irregularly occurring signals. The latter category is usually formed by artefacts.
61	EOG and muscle artifacts	Jung employed the infomax algorithm for the correction of EOG and muscle artifacts.
62	EOG	A comparison between ICA and PCA has been performed, concluding that ICA can effectively separate and remove contamination from a wide variety of artifact sources in EEG records with results comparing favorably to those obtained using PCA.
63	EOG	ICA has been applied for ERP data.
64	Eye movements, eye blinks, cardiac signals, muscle noise, and line noise	ICA has been extended for artifacts such as eye movements, eye blinks, cardiac signals, muscle noise, and line noise. The results were compared with PCA.
31	EOG	The clinical applications of ICA have been explored for 28 healthy controls and 22 clinical subjects while performing a visual event-related task.
65	EOG	ICA has been tested with noisy data and it was concluded that the performance of the methods were unaffected by an additive Gaussian noise source, as long as the signal-to-noise ratio remained above 50.

(Continued)

TABLE 2.6 (Continued)

Study	Artifact Type	Method
66	Electrocardiogram (EKG), eye movements, 50 Hz interference, muscle artifacts	ICA was tested for artifacts in short EEG samples. The study concluded that the distortion of the interictal activity measured by correlation analysis was minimal.
67	Sleep-related artifacts	ICA was applied to remove sleep-related artifacts and it was concluded that ICA is a useful technique for the evaluation of these variables with clinical interest in different sleep stages.
55	EOG	An automatic ICA-based method was proposed.
68	Eye, muscle, 50-Hz, electrocardiogram (ECG), and electrode artifacts	ICA is useful to remove artifacts from ictal recordings. When applied to ictal recordings, it increases the quality of the recording.
69	EOG	ICA is efficient at subtracting eye-movement artifacts, while retaining the EEG slow waves and making interpretation easier.
70	EOG	ICA-based automated method that does not require the availability of periocular EOG electrodes (SOBI).
16	Muscle artifacts	ICA has been extended for removing muscle activity from sleep EEG recordings. Comparison was performed for AMUSE, SOBI, infomax, and JADE.
71	EOG	Eye movement artifact removal was performed by ICA and compared with Gratton and Coles' method. ICA successfully identified components representing eye movement artifacts.
72	ECG, pulsation, respiration	Automated artifact removal as preprocessing refines neonatal seizure detection.

of ICA components could improve the situation. In this context, Winkler and colleagues have proposed an automatic selection based on a linear programming machine.[73] In addition, Chaumon et al. have provided a good account of the automatic selection of ICs after applying ICA.[74] The machine learning method provides subject independent removal of artifacts from EEG data. Table 2.7 provides a summary of all such methods that involve ICA with automatically subtracting components representing noise. For example, Yang et al. have proposed an automated method involving two stages.[75] In the first stage, simple ICA resulted in ICs. A kurtosis-based automatic method helps identify components that represent artifacts. Empirical model decomposition

TABLE 2.7 Automatic Identification of ICA Components

Study	Artifact Type	Method
76	ICA components that represent artifacts	Automatic selection of ICA components based on kurtosis criterion.
73	Artifactual source components from RT experiment	A combination of temporal decorrelation source separation (TDSEP) and a linear programming machine are used.
77	Artifacts because of head movements	The method employed an accelerometer to estimate head movements, which can be discarded using ICA-based decomposition.
78	EOG	The method uses modified multiscale sample entropy (mMSE) and kurtosis to automatically identify the independent eye blink artifactual components, and subsequently denoise these components using biorthogonal wavelet decomposition.
75	EOG	Spatial constraint independent component analysis based recursive least squares (SCICA-RLS) methods.
79	Physiological and nonbiological artifacts	(1) An event-related feature-based clustering algorithm used to identify artifacts that have physiological origins; and (2) electrode-scalp impedance information employed for identifying nonbiological artifacts

(EMD) allows for artifact components to be further separated, which in second stage are then used as the reference input to an adaptive filter in order to filter out the noise from EEG data. The second time ICA is applied EOG information is used; which is why it is known as constrained ICA.

EEG artifact reduction methods assume ICA as a standard method to separate artifacts and brain activity into independent components and employ ICA as a preprocessing step; this is known as ICA-based hybrid methods. These methods use ICA components to extract features or quantities that could help separate artifacts and normal EEG. For example, Delorme has computed higher components based on the ICA component.[76,80] Castellanos and colleagues have proposed wavelet enhanced ICA.[81] ICA has been integrated with dipole modeling as well.[82] However, the wavelet-based methods suffer from the difficulty of properly tuning the wavelet parameters. Table 2.8 provides a summary of different ICA-based hybrid methods.

2.7.8 Canonical Correlation Analysis

CCA is another BSS approach that has been used for the reduction of EEG-based artifacts. In this context, Table 2.9 has summarized different studies. However, most CCA-based methods only address muscle artifacts.

TABLE 2.8 ICA-Based Hybrid Methods for EEG-Based Artifact Reduction

Study	Artifact Type	Method
81	Muscle activity, eye blinks, and electrical noise	A wavelet threshold is applied to demixed, independent components as an intermediate step.
80	Muscle activity, eye blinks, and electrical noise	Higher order statistics and ICA
83	EOG and EMG	SVM and ICA
84	EOG	REG-ICA, an ICA-based method involving an adaptive filter, based on a stable version of the Recursive Least Square (sRLS) algorithm is applied to isolate the artifact and the actual neural signal.
85	Muscle activity, eye blinks, and electrical noise	ICA and SVM
82	Eye movement artifacts	An automatic method involving ICA and a dipole model.
86	EOG	SVM and ICA
87	Muscle activity, eye blinks, and electrical noise, etc.	ICA and wavelet denoising
88	Muscle artifacts	ICA and spectral ratio (SR)
89	Muscle and eye activities	Automatic wavelet enhanced ICA (AWICA), (employing kurtosis and Renyi entropy)
90	Muscle artifacts	ICA and state space modeling
77	Head movements	Accelerometer and ICA
91	Muscle and eye activities	Enhanced AWICA
92	EOG	ICA and system identification [Auto-Regressive eXogenous (ARX)]
93	Eye activity	Wavelet neural network and ICA
94	EOG	FASTICA and discrete wavelet transform (DWT)
95	Head movement	Accelerometer and ICA

(Continued)

TABLE 2.8 (Continued)

Study	Artifact Type	Method
96	Muscle activity, eye blinks, and electrical noise	Automated artifact elimination using linear discriminant analysis (LDA) for classification of feature vectors extracted from ICA components via image processing algorithms.
75	EOG	ICA and RLS
97	EOG	Outlier detection (OD) and ICA
98	OA	ICA and adaptive noise cancellation (ANC)

TABLE 2.9 CCA-Based Methods for EEG-Based Artifact Reduction

Study	Artifact Type	Method
99	Muscle artifact	CCA
100	Muscle artifact	CCA
101	Muscle artifact	CCA
102	Muscle artifact	CCA
103	Eye activity	EMD-CCA method
104	Muscle artifact	CCA

2.7.9 Adaptive Noise Cancellation

Adaptive noise cancellation (ANC) is a common process that employs adaptive filtering concepts. Its applications for EEG-based artifact reduction have been successful. Despite the need of separate reference channels, ANC provides online artifact reduction that is suitable for BCI applications and epileptic monitoring. Various researchers have proposed ANC-based EEG artifact reduction methods. Table 2.10 provides a summary of these methods.

2.7.10 Wavelet-Based Artifact Reduction/Thresholding Methods

EEG data are composite in nature and wavelet decomposition is suitable for analyzing EEG at multiple levels. After wavelet decomposition, EEG data can be divided into detailed and approximate components at various levels. Each level represents a certain frequency band. Wavelet-based methods mainly suffer due to manual selection of a

TABLE 2.10 ANC for EEG-Based Artifact Reduction

Methods	Artifact Type	Method
105	EOG	A neural network-based adaptive filtering method was designed. A real-time recurrent learning algorithm was employed for training the proposed neural network for fast convergence.
106	EOG	The method uses vertical EOG and horizontal EOG signals as two reference inputs. Each reference input is first processed by an FIR filter and then subtracted from the original EEG
107	EOG	The reference signal was then generated by the data collected from the forehead electrode being added to the data recorded from the temple electrode. The reference signal was also contaminated by EEG. To reduce the EEG interference, the reference signal was first low-pass filtered by a moving averaged filter and then applied to the ANC.
108	Line interference, EOG, and EEG	A cascade of three adaptive filters based on a Least Mean Square (LMS) algorithm is proposed. The first one eliminates line interference, the second adaptive filter removes the ECG artifacts, and the last one cancels EOG spikes.
109	Ocular and facial muscle artifacts	The proposed method uses horizontal EOG (HEOG), vertical EOG (VEOG), and EMG signals as three reference digital filter inputs. Real-time artifact removal is implemented by a multichannel Least Mean Square algorithm.
110	EOG	This study presents a novel ocular-artifact removal method based on adaptive filtering using reference signals from ocular SCs, which avoids the need for parallel EOG recordings.
111	Ocular and muscular artifacts	This generalized method consists of dividing the signal into several time-frequency windows, and in applying different spatial filters to each. Two steps are required to define one of these spatial filters: the first step consists of defining artifact spatial projection using the Common Spatial Pattern (CSP) method and the second consists of defining EEG spatial projection via regression.
112	Ocular, muscular, and cardiac artifacts	A new adaptive filter that involves a radial basis function as a neural network.
113	Head movements	A combination of band-pass filtering and multichannel adaptive filtering (AF); suitable for real-time MAR.

(Continued)

TABLE 2.10 (Continued)

Methods	Artifact Type	Method
114	EOG	An adaptive method that is based on a combination of filter banks and an eye tracker.
115	Eye blink	The proposed method adopts the Savitzky–Golay (SG) filter to extract the blink components from noisy EEG signals. The extracted component is then employed in an adaptive filter as a reference input. An adaptive neural fuzzy inference system (ANFIS) algorithm was implemented in adaptive filtering for the blink removal process.
116	Eye blinks, eye motions, amplitude drifts, and recording biases	ANC based on H^{∞} filtering.

suitable threshold, which renders the process subjective to the underlying data. On the contrary, automatic thresholding methods have been successfully employed for artifact reduction. Objective thresholding of wavelet coefficients provides automation to artifact correction methods. Table 2.11 shows a summary of these methods.

2.7.11 Template Matching Methods (Dynamic Time Warping and Empirical Model Decomposition)

Template matching methods are based on prior knowledge of the topography of the distribution of artifacts. The template for artifacts can be modeled theoretically or can be learned based on underlying EEG data. However, template matching methods can be subjective and a function of underlying EEG data. Table 2.12 provides a summary of all such methods.

2.7.12 Gait-Related Motion Artifacts

The reduction of motion-related artifacts has been considered differently than other types of artifacts, such as EOG or EMG. More recently, emphasis has been placed on motion artifacts, such as EEG recordings performed during exercise and jogging. Motion-related artifacts pose a different scenario for artifact reduction. In this context many studies have been performed. For example, in a study, the authors concluded that, the subjects' motion did not significantly affect their EEG during treadmill walking although precaution should

TABLE 2.11 Thresholding Methods for EEG-Based Artifact Reduction

Study	Artifact Type	Method
117	EOG	(1) A stationary wavelet transform is applied to the corrupted EEG; (2) thresholding of coefficients in the lower frequency bands is performed; and (3) a denoised signal is reconstructed.
118	EOG	The method is based on an appropriate threshold limit calculated from the statistical averages of the contaminated EEG signal and thresholding function. However, the threshold limit is empirically selected and is nonadaptive.
119	EOG	Automatically identifying slow varying OA zones and applying a wavelet-based adaptive thresholding algorithm only to the identified OA zones, this avoids the removal of background EEG information.
120	EOG	A statistical method, i.e., coefficient of variation for removing ocular artifacts from EEG recordings through wavelet transform.
121	EOG	An efficient combination of ICA, mutual information, and wavelet analysis for fully automatic ocular artifact suppression.
122	EOG	The proposed method is primarily based on stationary wavelet transform and takes the spectral band of seizure activities (i.e., 0.5–29 Hz) into account to separate artifacts from seizures

TABLE 2.12 Template Matching Methods for EEG-Based Artifact Reduction

Study	Artifact Type	Method
123	Eye blink	The method uses a fixed template for singling out the eye blink component after ICA.
124		The method requires the user to specify the minimum and maximum firing rates of the neurons. The algorithm iteratively estimates the morphology of the most prominent action potentials.
125	EOG	The method is based on linear predictive coding cepstrum (LPCC): coefficients of EOG pulses are extracted as feature vectors, which are used for eye movement pattern matching.
126	Eye blink	The method involves three steps: (1) an iterative process in which blink-events are detected and the blink-artifact waveform of the analyzed subject is estimated; (2) the generation of a signal modeling blink-artifact; and (3) suppression of this signal from the raw EEG.

be taken when gait speeds approach 4.5 km/h.[127] Overall, these findings suggest how MoBI methods may be safely deployed in neural, cognitive, and rehabilitation engineering applications.[127] Table 2.13 provides a summary of these methods.

TABLE 2.13 EEG Artifact Suppression Involving Motion-Related Artifacts

Study	Artifact Type	Method
128	Artifacts due to walking and running	This method uses stride time warping to remove gait artifacts from high-density EEG recorded during a visual oddball discrimination task performed while walking and running. Next, ICA is applied to parse the channel-based noise reduced EEG signals into maximally independent components (ICs) and then component-based template regression is performed.
129	Artifacts due to walking and running	Techniques to detect the presence of motion artifacts in PPGs given higher order statistical information present in the data. These data are analyzed in the time and frequency domains (FDs) and metrics to distinguish between clean and motion-corrupted data are identified.
130	Motion artifacts	A multichannel linear prediction filter.
131	Motion artifacts	The method has quantified the similarity between movement artifacts recorded by EEG electrodes and a head-mounted accelerometer. The aim is to isolate and record movement artifacts with EEG electrodes during walking. The electrophysiological signals are blocked using a nonconductive layer (silicone swim cap) and an electrically conductive scalp is simulated on top of the swim cap using a wig coated with conductive gel.
132	Gait-related movement artifacts	ICA and dipole fittings are used to localize movement-related component sources.
127	Gait-related movement artifacts	The authors investigated the potential contributions of motion artifacts in scalp EEG during treadmill walking at three different speeds (1.5, 3.0, and 4.5 km/h) using a wireless 64 channel active EEG system and a wireless inertial sensor attached to the subject's head.
133	Head movements	A phantom head device was used to mimic electrical properties of the human head with three controlled dipolar sources of electrical activity embedded in the phantom. The sinusoidal vertical motions were induced on the phantom head using a custom-built platform and the EEG signals were recorded with three different acquisition systems while the head was both stationary and in varied motion conditions.

(*Continued*)

TABLE 2.13 (Continued)

Study	Artifact Type	Method
17	Gait-related movement artifacts	A template correlation rejection as a novel method for identifying and rejecting EEG channels and independent components carrying motion-related artifacts. EEG data from 10 subjects were recorded during treadmill walking. The template correlation rejection method consists of creating templates of amplitude patterns and determining the fraction of total epochs presenting relevant correlation to the template.

TABLE 2.14 EEG Artifact Suppression Involving Subspace-Based Methods

Study	Artifact Type	Method
134	EOG, muscle, etc.	Generalized singular-value decomposition is used to separate multichannel electroencephalogram (EEG) into components found by optimizing a signal-to-noise quotient. These components are used to filter out artifacts.
135	Eye blinks, saccades, and other eye movements	Based on the concept that all ocular artifact components exist in a signal component subspace, the method can uniformly handle all types of ocular artifacts, including eye blinks, saccades, and other eye movements, by automatically identifying ocular components from decomposed signal components.
136	EOG	A deflation algorithm based on generalized eigenvalue decomposition for separating desired and undesired signal subspaces for multichannel EEG data.

2.7.13 Subspace-Based Methods

Subspace-based methods are based on the concept that noise and actual neuronal data can be discriminated into different subspaces. As a result, the noise subspace can be rejected while constructing the EEG signal. Table 2.14 provides a summary of subspace-based methods.

2.7.14 Hybrid Methods for Electroencephalography-Based Artifact Reduction

Recently, the trend for EEG-based artifact reduction has been transformed from signal methods to hybrid methods. In this context, a detailed description of each method is impossible; still a brief description is provided in Table 2.15.

2.7.15 Single Channel-Based Separation

Recently, single-channel EEG devices have become attractive because of their usability for measurement and their portability. The use of only one channel correction could be possible for BCI applications. Table 2.16 provides EEG-based methods involving a single EEG channel.

TABLE 2.15 Hybrid Methods for EEG-Based Artifact Reduction

Study	Artifact Type	Method
137	Eye and head movement	Machine learning and eye tracker (An eye tracker and motion sensor were also used to measure and provide the ground truth for the classification experiments.)
138	Subject generated artifacts	An AR model (parameters as features) and SVM classifier were used to discriminate among artifact conditions using the AR model parameters as features.
139	EOG	Wavelet neural network (combines the universal approximation characteristics of neural networks and the time/frequency property of wavelet transform).
140	EOG	High speed eye tracker for removing eye movement and blinks and adaptive filters (RLS and H^) to remove OA.
141	Eye blink	EOG Modeling and Kalman filtering to estimate the true EEG.
142	Muscle artifact	WPT-EMD and WPT-ICA algorithms
143	EOG	Neural network and adaptive fuzzy inference system
144	EOG	Complete Empirical Mode Decomposition with Adaptive Noise (CEEMDAN) and Renyi entropy (RE) is proposed for minimizing OAs from corrupted EEG signals.
145	Eye movement	The gaze signals, recorded by an eye tracker, share a similar temporal structure with the artifacts induced in EEG recordings by ocular movements. The proposed approach consists of estimating this specific common structure using multiple measurement vectors which are then used to denoise the EEG data
146	Eye and mucle artifacts	A combination of temporal motifs and dynamic time warping (DTW) methods is used to identify specific types of artifact segments quickly.

TABLE 2.16 Single Channel-Based Methods for EEG Artifact Reduction

Study	Artifact Type	Method
147	ECG	The method involves a two-step process: first, ECG artifact detection using the energy interval histogram method and then ECG artifact elimination using a modification of ensemble average subtraction.
148	Eye-blink artifact	Two-step nonnegative matrix factorization for single-channel EEG signals recorded at Fp1.
149	EOG	This method is based on a set of information on brain wave frequencies.
150	Eye blink	The method is based on combining digital filters with a rule-based decision system.
151	EOG	Unsupervised WT decomposition was systematically evaluated for the effectiveness of OA removal for a single-channel EEG system. Two commonly used WT methods; discrete wavelet transform (DWT) and stationary wavelet transform, were applied. Four WT basis functions, namely, haar, coif3, sym3, and bior4.4, were considered for OA removal with universal threshold and statistical threshold (ST).
152	EOG and EMG	The algorithm involves wavelet decomposition of original EEG into coefficients. Next, a surrogate time series is generated based on original EEG signal and a threshold is computed and compared with the coefficient values. Only the coefficients greater than the threshold value are used to reconstruct the EEG signal that is found clean from EOG and EMG.
153	Steady-state visual evoked potential (SSVEP)	STFT was employed to construct time-frequency images from raw EEG. Then SVD-based dimensionality reduction was performed to denoise the EEG data.
154	EOG	Maximum likelihood estimation is used to estimate EOG, which is further subtracted from the raw EEG.
155	EOG	A denoising technique is applied only to the OA zone to keep the critical neural information intact. The algorithm first detects the eye blinks (OA zone) using an algebraic approach, and then removes the artifacts from the OA zone using the discrete wavelet transform (DWT) decomposition method.
156	Muscle artifacts	This method is a combination of the ensemble empirical mode decomposition (EEMD) and joint blind source separation (JBSS) techniques.
157	EOG	Singular spectrum analysis (SSA) and adaptive noise canceler (ANC) were used to remove the EOG artifact from the contaminated EEG signal.

TABLE 2.17 Miscellaneous Methods for EEG-Based Artifact Reduction

Study	Artifact Type	Method
158	Eye blink	Extreme value statistics with a *p*-value acting as a threshold parameter.
PureEEG (159)	Artifacts due to lose electrode connections and machine interferences	It is a model-based method that performs estimations of artifacts, and EEG data based on the minimum mean square method.

2.7.16 Miscellaneous Methods

In this section, two studies have been described. In the first study, the authors claim that the eye-blink artifacts were removed successfully based on computing signal statistics and utilized the *p*-values as thresholding parameters.[158] The second study has addressed artifacts other than the EOG or EMG. The method has addressed artifacts caused by lose electrode connections and machine interferences. The proposed method employed the minimum mean square method. Table 2.17 provides a summary of these methods.

2.8 SUMMARY

This chapter provides a detailed description of EEG fundamentals, including a basic introduction to EEG data, EEG recording techniques, and the importance of EEG noise removal. Finally, the chapter discusses different methods for EEG preprocessing. In this chapter it is assumed that the reader is completely naïve to EEG and the details are explained accordingly.

References

1. Teplan M. Fundamentals of EEG measurement. *Meas Sci Rev*. 2002;2:1–11.
2. Proudfit GH, Bress JN, Foti D, Kujawa A, Klein DN. Depression and event-related potentials: emotional disengagement and reward insensitivity. *Curr Opin Psychol*. 2015;4:110–113.
3. Klem GH, Lüders HO, Jasper H, Elger C. The ten-twenty electrode system of the International Federation. *Electroencephalogr Clin Neurophysiol*. 1999;52:3.
4. Junghöfer M, Elbert T, Tucker DM, Braun C. The polar average reference effect: a bias in estimating the head surface integral in EEG recording. *Clin Neurophysiol*. 1999;110:1149–1155.
5. Yao D. Is the surface potential integral of a dipole in a volume conductor always zero? A cloud over the average reference of EEG and ERP. *Brain Topogr*. 2017;30:161–171.

6. Yao D, Wang L, Oostenveld R, Nielsen KD, Arendt-Nielsen L, Chen AC. A comparative study of different references for EEG spectral mapping: the issue of the neutral reference and the use of the infinity reference. *Physiol Meas*. 2005;26:173.
7. Tian Y, Yao D. Why do we need to use a zero reference? Reference influences on the ERPs of audiovisual effects. *Psychophysiology*. 2013;50:1282–1290.
8. Chella F, Pizzella V, Zappasodi F, Marzetti L. Impact of the reference choice on scalp EEG connectivity estimation. *J Neural Eng*. 2016;13:036016.
9. Trujillo LT, Stanfield CT, Vela RD. The effect of electroencephalogram (EEG) reference choice on information-theoretic measures of the complexity and integration of EEG signals. *Front Neurosci*. 2017;11:425.
10. Lei X, Liao K. Understanding the influences of EEG reference: a large-scale brain network perspective. *Front Neurosci*. 2017;11:205.
11. Yang P, Fan C, Wang M, Li L. A comparative study of average, linked mastoid, and REST references for ERP components acquired during fMRI. *Front Neurosci*. 2017;11:247.
12. Qin Y, Xin X, Zhu H, et al. A comparative study on the dynamic EEG center of mass with different references. *Front Neurosci*. 2017;11:509.
13. Hu S, Lai Y, Valdés-Sosa PA, Brings-Vega ML, Yao D. How do the reference montage and electrodes setup affect the measured scalp EEG potentials? *J Neural Eng*. 2017;15.
14. Gasser T, Ziegler P, Gattaz WF. The deleterious effect of ocular artefacts on the quantitative EEG, and a remedy. *Eur Arch Psychiatry Clin Neurosci*. 1992;241:352–356.
15. Hallez H, Vergult A, Phlypo R, et al. Muscle and eye movement artifact removal prior to EEG source localization. In: *Engineering in Medicine and Biology Society, 2006. EMBS'06. 28th Annual International Conference of the IEEE*; 2006:1002–1005.
16. Crespo-Garcia M, Atienza M, Cantero JL. Muscle artifact removal from human sleep EEG by using independent component analysis. *Ann Biomed Eng*. 2008;36:467–475.
17. Oliveira AS, Schlink BR, Hairston WD, König P, Ferris DP. A channel rejection method for attenuating motion-related artifacts in EEG recordings during walking. *Front Neurosci*. 2017;11.
18. Hillyard SA, Galambos R. Eye movement artifact in the CNV. *Electroencephalogr Clin Neurophysiol*. 1970;28:173–182.
19. Corby JC, Kopell BS. Differential contributions of blinks and vertical eye movements as artifacts in EEG recording. *Psychophysiology*. 1972;9:640–644.
20. McCallum W, Walter WG. The effects of attention and distraction on the contingent negative variation in normal and neurotic subjects. *Electroencephalogr Clin neurophysiol*. 1968;25:319–329.
21. Girton D, Kamiya J. A simple on-line technique for removing eye movement artifacts from the EEG. *Electroencephalogr Clin Neurophysiol*. 1973;34:212–216.
22. Barlow J, Remond A. Eye movement artifact nulling in EEGs by multichannel on-line EOG subtraction. *Electroencephalogr Clin Neurophysiol*. 1981;52:418–423.
23. Fortgens C, Bruin M De. Removal of eye movement and ECG artifacts from the noncephalic reference EEG. *Electroencephalogr Clin Neurophysiol*. 1983;56:90–96.
24. Gotman J, Skuce DR, Thompson CJ, Gloor P, Ives JR, Ray WF. Clinical applications of spectral analysis and extraction of features from electroencephalograms with slow waves in adult patients. *Electroencephalogr Clin Neurophysiol*. 1973;35:225–235.
25. Gotman J, Ives J, Gloor P. Frequency content of EEG and EMG at seizure onset: possibility of removal of EMG artefact by digital filtering. *Electroencephalogr Clinical Neurophysiol*. 1981;52:626–639.
26. Barlow JS. EMG artifact minimization during clinical EEG recordings by special analog filtering. *Electroencephalogr Clin Neurophysiol*. 1984;58:161–174.
27. Ives J, Schomer D. A 6-pole filter for improving the readability of muscle contaminated EEGs. *Electroencephalogr Clin Neurophysiol*. 1988;69:486–490.

28. Quilter P, MacGillivray B, Wadbrook D. The removal of eye movement artefact from EEG signals using correlation techniques. In: *Random Signal Analysis, IEEE Conference Publication*; 1977:93–100.
29. Whitton JL, Lue F, Moldofsky H. A spectral method for removing eye movement artifacts from the EEG. *Electroencephalogr Clin Neurophysiol*. 1978;44:735–741.
30. Gasser T, Möcks J. Correction of EOG artifacts in event-related potentials of the EEG: aspects of reliability and validity. *Psychophysiology*. 1982;19:472–480.
31. Jung T-P, Makeig S, Westerfield M, Townsend J, Courchesne E, Sejnowski TJ. Removal of eye activity artifacts from visual event-related potentials in normal and clinical subjects. *Clin Neurophysiol*. 2000;111:1745–1758.
32. Hagemann D, Naumann E. The effects of ocular artifacts on (lateralized) broadband power in the EEG. *Clin Neurophysiol*. 2001;112:215–231.
33. Gratton G, Coles MG, Donchin E. A new method for off-line removal of ocular artifact. *Electroencephalogr Clin Neurophysiol*. 1983;55:468–484.
34. Woestenburg J, Verbaten M, Slangen J. The removal of the eye-movement artifact from the EEG by regression analysis in the frequency domain. *Biol Psychol*. 1983;16:127–147.
35. Elbert T, Lutzenberger W, Rockstroh B, Birbaumer N. Removal of ocular artifacts from the EEG—a biophysical approach to the EOG. *Electroencephalogr Clin Neurophysiol*. 1985;60:455–463.
36. Gasser T, Sroka L, Möcks J. The correction of EOG artifacts by frequency dependent and frequency independent methods. *Psychophysiology*. 1986;23:704–712.
37. Semlitsch HV, Anderer P, Schuster P, Presslich O. A solution for reliable and valid reduction of ocular artifacts, applied to the P300 ERP. *Psychophysiology*. 1986;23:695–703.
38. Blogg T, Reid W. A digital technique for stimulus artifact reduction. *Electroencephalogr Clin Neurophysiol*. 1990;76:557–561.
39. Kenemans JL, Molenaar P, Verbaten MN, Slangen JL. Removal of the ocular artifact from the EEG: a comparison of time and frequency domain methods with simulated and real data. *Psychophysiology*. 1991;28:114–121.
40. Croft RJ, Barry RJ. EOG correction of blinks with saccade coefficients: a test and revision of the aligned-artefact average solution. *Clin Neurophysiol*. 2000;111:444–451.
41. Schlögl A, Keinrath C, Zimmermann D, Scherer R, Leeb R, Pfurtscheller G. A fully automated correction method of EOG artifacts in EEG recordings. *Clin Neurophysiol*. 2007;118:98–104.
42. Berg P, Scherg M. Dipole modelling of eye activity and its application to the removal of eye artefacts from the EEG and MEG. *Clin Phys Physiol Meas*. 1991;12:49.
43. Lins OG, Picton TW, Berg P, Scherg M. Ocular artifacts in EEG and event-related potentials I: scalp topography. *Brain Topogr*. 1993;6:51–63.
44. Lins OG, Picton TW, Berg P, Scherg M. Ocular artifacts in recording EEGs and event-related potentials II: source dipoles and source components. *Brain Topogr*. 1993;6:65–78.
45. Berg P, Scherg M. A multiple source approach to the correction of eye artifacts. *Electroencephalogr Clin Neurophysiol*. 1994;90:229–241.
46. Ille N. *Artifact Correction in Continuous Recordings of the Electro-and Magnetoencephalogram by Spatial Filtering*. Universität Mannheim; 2001.
47. Ille N, Berg P, Scherg M. Artifact correction of the ongoing EEG using spatial filters based on artifact and brain signal topographies. *J Clin Neurophysiol*. 2002;19:113–124.
48. Fitzgibbon SP, Powers DM, Pope KJ, Clark CR. Removal of EEG noise and artifact using blind source separation. *J Clin Neurophysiol*. 2007;24:232–243.
49. Lagerlund TD, Sharbrough FW, Busacker NE. Spatial filtering of multichannel electroencephalographic recordings through principal component analysis by singular value decomposition. *J Clin Neurophysiol*. 1997;14:73–82.

50. Makeig S, Jung T-P, Bell AJ, Ghahremani D, Sejnowski TJ. Blind separation of auditory event-related brain responses into independent components. *Proc Natl Acad Sci.* 1997;94:10979–10984.
51. Casarotto S, Bianchi AM, Cerutti S, Chiarenza GA. Principal component analysis for reduction of ocular artefacts in event-related potentials of normal and dyslexic children. *Clin Neurophysiol.* 2004;115:609–619.
52. Tomasevic L, Giambattistelli F, Pellegrino G, Rossini P, Siebner H, Tecchio F. P086 A data-driven method for TMS pulse artefact reduction from EEG data: sPCA. *Clin Neurophysiol.* 2017;128:e50.
53. Makeig S, Bell AJ, Jung T-P, Sejnowski TJ. Independent component analysis of electroencephalographic data. *Adv Neural Inf Process Syst.* 1996;8:145–151.
54. Bell AJ, Sejnowski TJ. An information-maximization approach to blind separation and blind deconvolution. *Neural Comput.* 1995;7:1129–1159.
55. Joyce CA, Gorodnitsky IF, Kutas M. Automatic removal of eye movement and blink artifacts from EEG data using blind component separation. *Psychophysiology.* 2004;41:313–325.
56. Kierkels JJ, van Boxtel GJ, Vogten LL. A model-based objective evaluation of eye movement correction in EEG recordings. *IEEE Trans Biomed Eng.* 2006;53:246–253.
57. Wallstrom GL, Kass RE, Miller A, Cohn JF, Fox NA. Automatic correction of ocular artifacts in the EEG: a comparison of regression-based and component-based methods. *Int J Psychophysiol.* 2004;53:105–119.
58. Romero S, Mañanas MA, Barbanoj MJ. A comparative study of automatic techniques for ocular artifact reduction in spontaneous EEG signals based on clinical target variables: a simulation case. *Comput Biol Med.* 2008;38:348–360.
59. Makeig S, Jung T-P, Ghahremani D, Sejnowski TJ. *Independent Component Analysis of Simulated ERP Data.* Institute for Neural Computation, University of California: technical report INC-9606; 1996.
60. Vigário RN. Extraction of ocular artefacts from EEG using independent component analysis. *Electroencephalogr Clin Neurophysiol.* 1997;103:395–404.
61. Jung T-P, Humphries C, Lee T-W, et al. Extended ICA removes artifacts from electroencephalographic recordings. In: *Advances in Neural Information Processing Systems;* 1998:894–900.
62. Jung T-P, Humphries C, Lee T-W, et al. Removing electroencephalographic artifacts: comparison between ICA and PCA. In: *Neural Networks for Signal Processing VIII, 1998. Proceedings of the 1998 IEEE Signal Processing Society Workshop;* 1998:63–72.
63. Jung T-P, Makeig S, Bell AJ, Sejnowski TJ. *Independent component analysis of electroencephalographic and event-related potential data. Central Auditory Processing and Neural Modeling.* Springer; 1998:189–197.
64. Jung T-P, Makeig S, Humphries C, et al. Removing electroencephalographic artifacts by blind source separation. *Psychophysiology.* 2000;37:163–178.
65. Vigon L, Saatchi M, Mayhew J, Fernandes R. Quantitative evaluation of techniques for ocular artefact filtering of EEG waveforms. *IEE Proc Sci Meas Technol.* 2000;147:219–228.
66. Iriarte J, Urrestarazu E, Valencia M, et al. Independent component analysis as a tool to eliminate artifacts in EEG: a quantitative study. *J Clin Neurophysiol.* 2003;20:249–257.
67. Romero S, Mananas M, Clos S, Gimenez S, Barbanoj M. Reduction of EEG artifacts by ICA in different sleep stages. In: *Engineering in Medicine and Biology Society, 2003. Proceedings of the 25th Annual International Conference of the IEEE;* 2003:2675–2678.
68. Urrestarazu E, Iriarte J, Alegre M, Valencia M, Viteri C, Artieda J. Independent component analysis removing artifacts in ictal recordings. *Epilepsia.* 2004;45:1071–1078.
69. Zhou W, Gotman J. Removing eye-movement artifacts from the EEG during the intracarotid amobarbital procedure. *Epilepsia.* 2005;46:409–414.

70. Gómez-Herrero G, De Clercq W, Anwar H, et al. Automatic removal of ocular artifacts in the EEG without an EOG reference channel. In: *Signal Processing Symposium, 2006. NORSIG 2006. Proceedings of the 7th Nordic*; 2006:130–133.
71. Mennes M, Wouters H, Vanrumste B, Lagae L, Stiers P. Validation of ICA as a tool to remove eye movement artifacts from EEG/ERP. *Psychophysiology*. 2010;47:1142–1150.
72. De Vos M, Deburchgraeve W, Cherian P, et al. Automated artifact removal as preprocessing refines neonatal seizure detection. *Clin Neurophysiol*. 2011;122:2345–2354.
73. Winkler I, Haufe S, Tangermann M. Automatic classification of artifactual ICA-components for artifact removal in EEG signals. *Behav Brain Funct*. 2011;7:30.
74. Chaumon M, Bishop DV, Busch NA. A practical guide to the selection of independent components of the electroencephalogram for artifact correction. *J Neurosci Methods*. 2015;250:47–63.
75. Yang B-H, He L-F, Lin L, Wang Q. Fast removal of ocular artifacts from electroencephalogram signals using spatial constraint independent component analysis based recursive least squares in brain-computer interface. *Front Inform Technol El*. 2015;16:486–496.
76. Delorme A, Makeig S, Sejnowski T. Automatic artifact rejection for EEG data using high-order statistics and independent component analysis. In: *Proceedings of the Third International ICA Conference*; 2001:9–12.
77. Daly I, Billinger M, Scherer R, Müller-Putz G. On the automated removal of artifacts related to head movement from the EEG. *IEEE Trans Neural Syst Rehabil Eng*. 2013;21:427–434.
78. Mahajan R, Morshed BI. Unsupervised eye blink artifact denoising of EEG data with modified multiscale sample entropy, kurtosis, and Wavelet-ICA. *IEEE J Biomed Health Inform*. 2015;19:158–165.
79. Zou Y, Nathan V, Jafari R. Automatic identification of artifact-related independent components for artifact removal in EEG recordings. *IEEE J Biomed Health Inform*. 2016;20:73–81.
80. Delorme A, Sejnowski T, Makeig S. Enhanced detection of artifacts in EEG data using higher order statistics and independent component analysis. *Neuroimage*. 2007;34:1443–1449.
81. Castellanos NP, Makarov VA. Recovering EEG brain signals: artifact suppression with wavelet enhanced independent component analysis. *J Neurosci Methods*. 2006;158:300–312.
82. Zhou W, Gotman J. Automatic removal of eye movement artifacts from the EEG using ICA and the dipole model. *Pro Nat Sci*. 2009;19:1165–1170.
83. Halder S, Bensch M, Mellinger J, et al. Online artifact removal for brain-computer interfaces using support vector machines and blind source separation. *Comput Intell Neurosci*. 2007;2007.
84. Klados MA, Papadelis CL, Bamidis PD. REG-ICA: a new hybrid method for eog artifact rejection. In: *Information Technology and Applications in Biomedicine, 2009. ITAB 2009. 9th International Conference on*; 2009:1–4.
85. Shao S-Y, Shen K-Q, Ong CJ, Wilder-Smith EP, Li X-P. Automatic EEG artifact removal: a weighted support vector machine approach with error correction. *IEEE Trans Biomed Eng*. 2009;56:336–344.
86. Gao JF, Yang Y, Lin P, Wang P, Zheng CX. Automatic removal of eye-movement and blink artifacts from EEG signals. *Brain Topogr*. 2010;23:105–114.
87. Akhtar MT, Mitsuhashi W, James CJ. Employing spatially constrained ICA and wavelet denoising, for automatic removal of artifacts from multichannel EEG data. *Signal Process*. 2012;92:401–416.
88. Ma J, Tao P, Bayram S, Svetnik V. Muscle artifacts in multichannel EEG: characteristics and reduction. *Clin Neurophysiol*. 2012;123:1676–1686.

89. Mammone N, La Foresta F, Morabito FC. Automatic artifact rejection from multichannel scalp EEG by wavelet ICA. *IEEE Sens J*. 2012;12:533–542.
90. Santillán-Guzmán A, Heute U, Stephani U, Galka A. Muscle artifact suppression using independent-component analysis and state-space modeling. In: *Engineering in Medicine and Biology Society (EMBC), 2012 Annual International Conference of the IEEE*; 2012:6500–6503.
91. Mammone N, Morabito FC. Enhanced automatic wavelet independent component analysis for electroencephalographic artifact removal. *Entropy*. 2014;16:6553–6572.
92. Wang Z, Xu P, Liu T, Tian Y, Lei X, Yao D. Robust removal of ocular artifacts by combining independent component analysis and system identification. *Biomed Signal Process Control*. 2014;10:250–259.
93. Burger C, van den Heever DJ. Removal of EOG artefacts by combining wavelet neural network and independent component analysis. *Biomed Signal Process Control*. 2015;15:67–79.
94. Mingai L, Shuoda G, Guoyu Z, Yanjun S, Jinfu Y. Removing ocular artifacts from mixed EEG signals with FastKICA and DWT. *J Intell Fuzzy Syst*. 2015;28:2851–2861.
95. Onikura K, Katayama Y, Iramina K. Evaluation of a method of removing head movement artifact from EEG by independent component analysis and filtering. *Adv Biomed Eng*. 2015;4:67–72.
96. Radüntz T, Scouten J, Hochmuth O, Meffert B. EEG artifact elimination by extraction of ICA-component features using image processing algorithms. *J Neurosci Methods*. 2015;243:84–93.
97. Çınar S, Acır N. A novel system for automatic removal of ocular artefacts in EEG by using outlier detection methods and independent component analysis. *Expert Syst Appl*. 2017;68:36–44.
98. Jafarifarmand A, Badamchizadeh M-A, Khanmohammadi S, Nazari MA, Tazehkand BM. Real-time ocular artifacts removal of EEG data using a hybrid ICA-ANC approach. *Biomed Signal Process Control*. 2017;31:199–210.
99. Gao J, Zheng C, Wang P. Online removal of muscle artifact from electroencephalogram signals based on canonical correlation analysis. *Clin EEG Neurosci*. 2010;41:53–59.
100. Vos DM, Riès S, Vanderperren K, et al. Removal of muscle artifacts from EEG recordings of spoken language production. *Neuroinformatics*. 2010;8:135–150.
101. De Clercq W, Vergult A, Vanrumste B, et al. A new muscle artifact removal technique to improve the interpretation of the ictal scalp electroencephalogram. In: *Engineering in Medicine and Biology Society, 2005. IEEE-EMBS 2005. 27th Annual International Conference of*; 2006:944–947.
102. De Clercq W, Vergult A, Vanrumste B, Van Paesschen W, Van Huffel S. Canonical correlation analysis applied to remove muscle artifacts from the electroencephalogram. *IEEE Trans Biomed Eng*. 2006;53:2583–2587.
103. Soomro MH, Badruddin N, Yusoff MZ, Jatoi MA. Automatic eye-blink artifact removal method based on EMD-CCA. In: *Complex Medical Engineering (CME), 2013 ICME International Conference on*; 2013:186–190.
104. Vergult A, De Clercq W, Palmini A, et al. Improving the interpretation of ictal scalp EEG: BSS–CCA algorithm for muscle artifact removal. *Epilepsia*. 2007;48:950–958.
105. Selvan S, Srinivasan R. Removal of ocular artifacts from EEG using an efficient neural network based adaptive filtering technique. *IEEE Signal Process Lett*. 1999;6:330–332.
106. He P, Wilson G, Russell C. Removal of ocular artifacts from electro-encephalogram by adaptive filtering. *Med Biol Eng Comput*. 2004;42:407–412.
107. Erfanian A, Mahmoudi B. Real-time ocular artifact suppression using recurrent neural network for electro-encephalogram based brain-computer interface. *Med Biol Eng Comput*. 2005;43:296–305.

108. Correa AG, Laciar E, Patino H, Valentinuzzi M. Artifact removal from EEG signals using adaptive filters in cascade. *J Phys: Conf Ser*. 2007;90:012081.
109. Mehrkanoon S, Moghavvemi M, Fariborzi H. Real time ocular and facial muscle artifacts removal from EEG signals using LMS adaptive algorithm. In: *Intelligent and Advanced Systems, 2007. ICIAS 2007. International Conference on*; 2007:1245–1250.
110. Chan H-L, Tsai Y-T, Meng L-F, Wu T. The removal of ocular artifacts from EEG signals using adaptive filters based on ocular source components. *Ann Biomed Eng*. 2010;38:3489–3499.
111. Boudet S, Peyrodie L, Forzy G, Pinti A, Toumi H, Gallois P. Improvements of adaptive filtering by optimal projection to filter different artifact types on long duration EEG recordings. *Comput Methods Programs Biomed*. 2012;108:234–249.
112. Jafarifarmand A, Badamchizadeh MA. Artifacts removal in EEG signal using a new neural network enhanced adaptive filter. *Neurocomputing*. 2013;103:222–231.
113. Mihajlović V, Patki S, Grundlehner B. The impact of head movements on EEG and contact impedance: an adaptive filtering solution for motion artifact reduction. In: *Engineering in Medicine and Biology Society (EMBC), 2014 36th Annual International Conference of the IEEE*; 2014:5064–5067.
114. Orhan U, Mathan S. Removal of ocular artifacts with the utilization of filter banks. In: *International Conference on Augmented Cognition*; 2015:504–513.
115. Rahman FA, Othman MF. Eye blinks removal in single-channel EEG using Savitzky-Golay referenced adaptive filtering: a comparison with independent component analysis method. *ARPN J Eng Appl Sci*. 2015;10:18147–18154.
116. Kilicarslan A, Grossman RG, Contreras-Vidal JL. A robust adaptive denoising framework for real-time artifact removal in scalp EEG measurements. *J Neural Eng*. 2016;13:026013.
117. Zikov T, Bibian S, Dumont GA, Huzmezan M, Ries C. A wavelet based de-noising technique for ocular artifact correction of the electroencephalogram. *Engineering in Medicine and Biology, 2002. 24th Annual Conference and the Annual Fall Meeting of the Biomedical Engineering Society EMBS/BMES Conference, 2002. Proceedings of the Second Joint*; 2002:98–105.
118. Krishnaveni V, Jayaraman S, Malmurugan N, Kandaswamy A, Ramadoss K. Non adaptive thresholding methods for correcting ocular artifacts in EEG. *Acad Open Internet J*. 2004;13:1–9.
119. Krishnaveni V, Jayaraman S, Anitha L, Ramadoss K. Removal of ocular artifacts from EEG using adaptive thresholding of wavelet coefficients. *J Neural Eng*. 2006;3:338.
120. Kumar PS, Arumuganathan R, Sivakumar K, Vimal C. Removal of ocular artifacts in the EEG through wavelet transform without using an EOG reference channel. *Int J Open Problems Compt Math*. 2008;1:188–200.
121. Ghandeharion H, Erfanian A. A fully automatic ocular artifact suppression from EEG data using higher order statistics: improved performance by wavelet analysis. *Med Eng Phys*. 2010;32:720–729.
122. Islam MK, Rastegarnia A, Yang Z. A wavelet-based artifact reduction from scalp eeg for epileptic seizure detection. *IEEE J Biomed Health Inform*. 2016;20:1321–1332.
123. Li Y, Ma Z, Lu W, Li Y. Automatic removal of the eye blink artifact from EEG using an ICA-based template matching approach. *Physiol Meas*. 2006;27:425.
124. Kim S, McNames J. Automatic spike detection based on adaptive template matching for extracellular neural recordings. *J Neurosci Methods*. 2007;165:165–174.
125. Lv Z, Wu X-P, Li M, Zhang D. A novel eye movement detection algorithm for EOG driven human computer interface. *Pattern Recognit Lett*. 2010;31:1041–1047.
126. Valderrama JT, de la Torre A, Van Dun B. An automatic algorithm for blink-artifact suppression based on iterative template matching: application to single channel recording of cortical auditory evoked potentials. *J Neural Eng*. 2017.

127. Nathan K, Contreras-Vidal JL. Negligible motion artifacts in scalp electroencephalography (EEG) during treadmill walking. *Front Hum Neurosci*. 2016;9:708.
128. Gwin JT, Gramann K, Makeig S, Ferris DP. Removal of movement artifact from high-density EEG recorded during walking and running. *J Neurophysiol*. 2010;103:3526–3534.
129. Krishnan R, Natarajan B, Warren S. Two-stage approach for detection and reduction of motion artifacts in photoplethysmographic data. *IEEE Trans Biomed Eng*. 2010;57:1867–1876.
130. Bertrand A, Mihajlovic V, Grundlehner B, Van Hoof C, Moonen M. Motion artifact reduction in EEG recordings using multi-channel contact impedance measurements. In: *Biomedical Circuits and Systems Conference (BioCAS), 2013 IEEE*; 2013:258–261.
131. Kline JE, Huang HJ, Snyder KL, Ferris DP. Isolating gait-related movement artifacts in electroencephalography during human walking. *J Neural Eng*. 2015;12:046022.
132. Snyder KL, Kline JE, Huang HJ, Ferris DP. Independent component analysis of gait-related movement artifact recorded using EEG electrodes during treadmill walking. *Front Hum Neurosci*. 2015;9.
133. Oliveira AS, Schlink BR, Hairston WD, König P, Ferris DP. Induction and separation of motion artifacts in EEG data using a mobile phantom head device. *J Neural Eng*. 2016;13:036014.
134. Anderson CW, Knight JN, O'Connor T, Kirby MJ, Sokolov A. Geometric subspace methods and time-delay embedding for EEG artifact removal and classification. *IEEE Trans Neural Syst Rehabil Eng*. 2006;14:142–146.
135. Ma J, Bayram S, Tao P, Svetnik V. High-throughput ocular artifact reduction in multichannel electroencephalography (EEG) using component subspace projection. *J Neurosci Methods*. 2011;196:131–140.
136. Sameni R, Gouy-Pailler C. An iterative subspace denoising algorithm for removing electroencephalogram ocular artifacts. *J Neurosci Methods*. 2014;225:97–105.
137. Chadwick NA, McMeekin DA, Tan T. Classifying eye and head movement artifacts in EEG signals. In: *Digital Ecosystems and Technologies Conference (DEST), 2011 Proceedings of the 5th IEEE International Conference on*; 2011:285–291.
138. Lawhern V, Hairston WD, McDowell K, Westerfield M, Robbins K. Detection and classification of subject-generated artifacts in EEG signals using autoregressive models. *J Neurosci Methods*. 2012;208:181–189.
139. Nguyen H-AT, Musson J, Li F, et al. EOG artifact removal using a wavelet neural network. *Neurocomputing*. 2012;97:374–389.
140. Noureddin B, Lawrence PD, Birch GE. Online removal of eye movement and blink EEG artifacts using a high-speed eye tracker. *IEEE Trans Biomed Eng*. 2012;59:2103–2110.
141. Shahabi H, Moghimi S, Zamiri-Jafarian H. EEG eye blink artifact removal by EOG modeling and Kalman filter. In: *Biomedical Engineering and Informatics (BMEI), 2012 5th International Conference on*; 2012:496–500.
142. Bono V, Jamal W, Das S, Maharatna K. Artifact reduction in multichannel pervasive EEG using hybrid WPT-ICA and WPT-EMD signal decomposition techniques. In: *Acoustics, Speech and Signal Processing (ICASSP), 2014 IEEE International Conference on*; 2014:5864–5868.
143. Hu J, Wang C-S, Wu M, Du Y-x, He Y, She J. Removal of EOG and EMG artifacts from EEG using combination of functional link neural network and adaptive neural fuzzy inference system. *Neurocomputing*. 2015;151:278–287.
144. Guarascio M, Puthusserypady S. Automatic minimization of ocular artifacts from electroencephalogram: a novel approach by combining complete EEMD with adaptive noise and Renyi's entropy. *Biomed Signal Process Control*. 2017;36:63–75.
145. Maurandi V, Rivet B, Phlypo R, Guérin–Dugué A, Jutten C. Multimodal approach to remove ocular artifacts from EEG signals using multiple measurement vectors. In: *International Conference on Latent Variable Analysis and Signal Separation*; 2017:563–573.

146. Shaw L, Routray A, Sanchay S. A robust motifs based artifacts removal technique from EEG. *Biomed Phys Eng Express*. 2017;3:035010.
147. Park H-J, Jeong D-U, Park K-S. Automated detection and elimination of periodic ECG artifacts in EEG using the energy interval histogram method. *IEEE Trans Biomed Eng*. 2002;49:1526–1533.
148. Kanoga S, Mitsukura Y. Eye-blink artifact reduction using 2-step nonnegative matrix factorization for single-channel electroencephalographic signals. *J Signal Process*. 2014;18:251–257.
149. Zammouri A, Aitmoussa A, Chevallier S, Monacelli E. Intelligentocular artifacts removal in a noninvasive singlechannel EEG recording. In: *Intelligent Systems and Computer Vision (ISCV), 2015*; 2015:1–5.
150. Chang W-D, Cha H-S, Kim K, Im C-H. Detection of eye blink artifacts from single prefrontal channel electroencephalogram. *Comput Methods Programs Biomed*. 2016;124:19–30.
151. Khatun S, Mahajan R, Morshed BI. Comparative study of wavelet-based unsupervised ocular artifact removal techniques for single-channel EEG data. *IEEE J Transl Eng Health Med*. 2016;4:1–8.
152. Chavez M, Grosselin F, Bussalb A, Fallani FDV, Navarro-Sune X. Surrogate-based artifact removal from single-channel EEG. *arXiv preprint arXiv:1704.07603*; 2017.
153. Wang Y, Xu G, Zhang S, Luo A, Li M, Han C. EEG signal co-channel interference suppression based on image dimensionality reduction and permutation entropy. *Signal Process*. 2017;134:113–122.
154. Karimi S, Molaee-Ardekani B, Shamsollahi MB, Leroy C, Derambure P. Automatic ocular correction in EEG recordings using maximum likelihood estimation. In: *Signal Processing and Information Technology (ISSPIT), 2013 IEEE International Symposium on*; 2013:000164–000169.
155. Majmudar CA, Mahajan R, Morshed BI. Real-time hybrid ocular artifact detection and removal for single channel EEG. In: *Electro/Information Technology (EIT), 2015 IEEE International Conference on*; 2015:330–334.
156. Chen X, Liu A, Peng H, Ward RK. A preliminary study of muscular artifact cancellation in single-channel EEG. *Sensors*. 2014;14:18370–18389.
157. Maddirala AK, Shaik RA. Removal of EOG Artifacts from single channel EEG signals using combined singular spectrum analysis and adaptive noise canceler. *IEEE Sensors J*. 2016;16:8279–8287.
158. Klein A, Skrandies W. A reliable statistical method to detect eyeblink-artefacts from electroencephalogram data only. *Brain Topogr*. 2013;26:558–568.
159. Hartmann M, Schindler K, Gebbink T, Gritsch G, Kluge T. PureEEG: automatic EEG artifact removal for epilepsy monitoring. *Neurophysiol Clin*. 2014;44:479–490.

CHAPTER

3

Electroencephalography-Based Brain Functional Connectivity and Clinical Implications

3.1 INTRODUCTION

Electroencephalography (EEG)-based brain connectivity analysis has shown important research findings. For example, analysis has shown evidence in the existence of synchronous activity among different brain structures during various cognitive and problem solving tasks (see e.g., Ref. 1). This synchronous activity may represent a successful exchange of information between different brain areas. Any abnormalities during synchronous activity may indicate abnormal brain connectivity. In general, brain connectivity can be defined in terms of anatomical or structural connectivity, FC, and effective connectivity (EC). Anatomical connectivity refers to the physical connections between different brain structures involving the linking sets of neurons or neuronal elements.[2] Anatomical connectivity could be studied with a modality such as diffusion tensor imaging (DTI). On the other hand, FC entails statistical dependence between signals recorded from spatially located sensors,[3] while EC refers to the causal relationship between them.[3,4] Both EC and FC can be studied with EEG, functional magnetic resonance imaging (fMRI),[5] magnetoencephalogram (MEG), positron emission tomography (PET), and static positron emission tomography (SPECT).

EEG as a standard modality can investigate the brain either during rest or nonrest conditions such as performance of a working memory task. As widely established, EEG data are composite in nature and includes various frequencies such as delta (0.1–4 Hz), theta (4–8 Hz), alpha (8–12 Hz), beta (12–20 Hz), and gamma (>20 Hz). These EEG data bands have shown association with different physiological states, for example, EEG alpha activity is considered as an index of relaxation

EEG-Based Experiment Design for Major Depressive Disorder
DOI: https://doi.org/10.1016/B978-0-12-817420-3.00003-5

and can be observed during eyes close (EC) or eyes open (EO) conditions in healthy controls. In addition, the beta band has been associated with mental activity. The theta band has been reported in healthy controls during sleep. On the contrary, deviation from normal EEG patterns could implicate a brain pathology. For example, alcoholics have shown less theta activity when compared to healthy controls.[6] Moreover, in comparison to healthy controls, MDD patients have shown alpha interhemispheric lateralization.[7] Furthermore, the precision and accuracy of these patterns could lead to the discovery of endophenotypes for a particular mental illness.[8]

Recently, various EEG-based research studies were performed for the assessment of brain connectivity.[9] Since EEG data have nonlinear dynamics, several types of methods are required in order to overcome the shortcomings and challenges faced by previous methods. Hence, advancement in the development of computationally intensive techniques becomes important. In broader sense, these EEG-based methods could be categorized as bivariate or multivariate methods.[10] Bivariate methods mainly involved computing the connectivity between two signals or scalp locations. On the contrary, multivariate methods involved the whole dataset (more than two signals simultaneously) to train statistical models.[11] Furthermore, various FC methods come in different varieties and categorized accordingly such as linear versus nonlinear methods and parametric versus nonparametric measures. In addition, these methods can be categorized into several different types such as classical measures, for example, correlation of EEG data between two scalp locations or its equivalent in frequency domain, that is, magnitude square coherence. Different EEG-based methods have been proposed for the assessment of brain FC, for example, interhemispheric coherence, phase lag index (PLI), mutual information (MI), SL, phase locking and phase coherence, correlation, and imaginary coherence. The choice of a method depends on the underlying assumptions and the type of analysis being performed.

Although EEG-based FC analyses show abnormalities associated with various mental illnesses, a structured review with the latest clinical updates on these measures is required. Therefore, the aim of this chapter is to provide information on various aspects of brain connectivity such as EEG-based quantitative methods, their clinical applications, and MATLAB toolboxes available for data analysis. Moreover, EEG-based FC analysis could be confounded due to volume conduction effects, signal to noise ratio issues, common reference problem, common input problem, or sample bias problem. A detailed description of these factors can be found elsewhere.[1] This chapter mainly focuses on the clinical findings of EEG-based FC methods. Sections 3.2.1–3.2.6 discuss the applications of these methods in the context of Alzheimer's, mild cognitive impairment (MCI), schizophrenia, depression, and alcoholism. In this chapter, Section 3.2 explains the clinical applications of FC measures and Section 3.3 illustrates the available open source tools used for measuring connectivity. Finally, the chapter is concluded with the conclusion section.

3.2 CLINICAL IMPLICATIONS OF ELECTROENCEPHALOGRAPHY-BASED BRAIN CONNECTIVITY METHODS

This section highlights brain connectivity measures in the context of various neurological disorders such as epilepsy, schizophrenia, MDD, and Alzheimer's disease. In the case of depression, a decreased FC has been observed when compared with healthy controls.[12] Other than depression, mental illnesses such as schizophrenia, epilepsy, and Alzheimer's have shown abnormal FC patterns when compared with healthy controls.[13] In a recent study, it was validated that the connections between the frontal and parietal regions are modified by anesthesia.[14] A detailed description of the implications of EEG-based FC differences is provided in the relevant sections.

3.2.1 Alzheimer's

Alzheimer's has been associated with brain cognitive impairments. Approximately 200,000 Americans under the age of 65 have younger-onset Alzheimer's disease (also known as early onset Alzheimer's). Unfortunately, Alzheimer's is the sixth leading cause of death in the United States.[15] In the literature, EEG-based FC measures such as coherence, phase locking index (PLI), SL, and MI have been utilized for discriminating between Alzheimer's patients and healthy controls.

3.2.1.1 Interhemispheric Coherence

According to the literature, most studies have reported decreased EEG coherence during Alzheimer's disease, suggesting decreased interaction between brain areas.[16,17] In another study, rhythmic coherences (particularly in the alpha and beta bands) and cumulative mutual information (CMI) variables as well as harmonic coherences (particularly related to 3 Hz PS) were reported to be significantly lower in probable AD than in control subjects.[18] Furthermore, in a study, global correlation dimension and coherence pointed to a decreased functional cortical connectivity.[19] Additionally, coherence in the alpha band was reported to increase in the control group following a given task, but was not seen in the AD group.[20] In short, there was increasing evidence of decreasing coherence in Alzheimer's patients compared to control subjects.[21–24] In a different study, FC measures (such as correlation coefficients, mean square and phase coherence, Granger causality, phase synchrony indices, information theoretic divergence measure, phase space-based measures, and stochastic event synchrony) were compared for the diagnosis of Alzheimer disease and Granger causality was found to be efficient as a diagnostic marker for Alzheimer's.[13] On the contrary, a study has concluded higher gamma coherence in Alzheimer's patients than in healthy controls.[25]

3.2.1.2 *Phase Lag Index*

Numerous studies have investigated disease severity utilizing FC measures. For example, the PLI has been utilized to study connectivity differences between Alzheimer's patients and control subjects.[26,27] A slightly different approach involved the quantification of FC with PLI that resulted into a connectivity matrix. The matirx was utilized to construct the brain network. Further, different network parameters were considered as features and used to study the differences in cognitive decline between study groups, including in Alzheimer's patients.[26] In a study, disease severity was investigated resulting in the conclusion that the FC decreases with increasing disease severity in the alpha band.[27]

3.2.1.3 *Synchronization Likelihood*

Some studies have utilized SL to study connectivity differences between healthy subjects and Alzheimer's patients.[28,29] Alzheimer's patients showed lower levels of synchronization in different frequency bands: alpha (10–13 Hz), beta (13–30 Hz),[28] α (10–12 Hz) and β (12–30 Hz).[29] MI provides statistical dependence between two time series and has been utilized to study connectivity differences between healthy subjects and Alzheimer's patients. According to a study, it was found to be lower in AD patients when compared with healthy controls.[30] Table 3.1 shows EEG-based studies that were conducted for Alzheimer disease and provides a summary of EEG-based FC measures and the datasets utilized for the Alzheimer's disease.

3.2.2 Mild Cognitive Impairment

Mild cognitive impairment (MCI) causes a slight but noticeable and measurable decline in cognitive abilities, including memory and thinking skills. A person with MCI is at an increased risk of developing Alzheimer's or another dementia. It is estimated that between 5% and 20% of people over the age of 65 have MCI.[31]

TABLE 3.1 EEG-Based Functional Connectivity (FC) Methods for Alzheimer's Disease

Year	EEG-Based FC Measure	Data Set Used	Finding
2017[25]	Coherence	Healthy = 21 Untreated AD = 21, treated AD = 18	Gamma coherences were significantly higher in AD patients as compared with healthy controls.
2016[26]	Phase lag index (PLI), weighted network, and minimum spanning trees	AD patients = 69, subjective cognitive decline (SCD) patients = 64	Occipital loss of network organization in AD patients was reported.
2015[27]	Phase lag index (PLI)	AD patients = 318, Healthy = 133	The AD patients exhibited decreased FC with increasing disease severity in the alpha band.
2014[16]	Coherence	AD patients = 15 Healthy = 15	Compared with the control group, the pairwise coherence of the AD group is significantly decreased, especially for the theta and alpha frequency bands in the frontal and parieto-occipital regions.
2013[18]	Spectral coherence and mutual information	AD patients = 16 Healthy = 15	The coherences were significantly lower in the probable AD than in normal subjects.
2010[13]	Granger causality and stochastic event synchrony	AD patients = 25 Healthy = 56	The discrimination between AD patients and healthy controls resulted in a classification accuracy of 83%.
2008[17]	Coherence	AD patients = 21 (10 untreated, 11 treated) Healthy = 19	The control group showed higher values of evoked coherence in the left frontoparietal electrode pair in the theta frequency band (P <.01) and higher values of evoked coherence in the right frontoparietal electrode pair in the delta band (P <.01) when compared to the treated AD group.
2008[19]	Global coherence and correlation dimension	AD patients = 15 Healthy = 21	In AD, global correlation dimension and coherence are changed especially in the higher frequency ranges, both pointing to decreased functional cortical connectivity.
2007[20]	EEG spectra and coherence	AD patients = 14 Healthy = 10	Coherence in this frequency band increased in the control group following the given task not seen in the AD group.

(Continued)

TABLE 3.1 (Continued)

Year	EEG-Based FC Measure	Data Set Used	Finding
2005[21]	Coherence	AD patients = 35 Healthy = 33	The general decrease of AD patients in inter- and intrahemispheric EEG coherence was more significant than that of the normal controls at resting EEG, with the most striking decrease observed in the alpha-1 (8.0–9.0 Hz) and alpha-2 (9.5–12.5 Hz) bands.
2005[28]	Synchronization likelihood	AD patients = 24 Healthy = 19	The mean level of EEG synchronization was lower in Alzheimer's patients in the upper alpha (10–13 Hz) and beta (13–30 Hz) bands. Spontaneous fluctuations of synchronization were diminished in Alzheimer's patients in the lower alpha (8–10 Hz) and beta bands.
2004[29]	Synchronization likelihood	AD patients = 14 MCI patients = 11 Healthy = 14	The SL was significantly decreased in the upper alpha (10–12 Hz) and beta (12–30 Hz) bands in AD compared to persons with subjective memory complaints.
2003[22]	Coherence	AD patients = 10 Healthy = 10	When compared with normal controls, AD patients had reduced upper alpha coherence between the central and right temporal cortex.
2003[23]	Coherence	AD patients = 10 Healthy = 10	This study suggests that coherence is a useful tool for understanding electrophysiologic change of AD and for correlating with the severity of cognitive dysfunction. These coherence changes can be interpreted as the effects of neuronal loss and neocortical disconnection.
2001[30]	Mutual information	AD patients = 15 Healthy = 15	The local CMI in AD subjects was lower than that in normal controls, especially in the frontal and anterotemporal regions.
2000[24]	Coherence	AD patients = 11 Healthy = 5	A decreased coherence at the occipital in the Alzheimer's group for both alpha sub-bands was reported.

3.2.2.1 *Interhemispheric Coherence*

Interhemispheric coherence has been employed for the evaluation of EEG data from MCI patients.[32–34] Studies have reported an increased coherence in MCI groups during working memory tasks. However, the same difference cannot be observed during rest state.[32] In a study, EEG has been investigated as a predictor of developing Alzheimer disease from MCI.[33] The study found that frontoparietal midline coherence as well as delta (temporal), theta (parietal, occipital, and temporal), and alpha-1 (central, parietal, occipital, temporal, and limbic) sources were stronger in MCI Converted than in stable subjects ($P < .05$). Similarly, MCI patients who showed an inclination for Alzheimer disease have exhibited baseline alpha band temporoparietal coherence.[34] In a study, patients with mild cognitive impairment (MCI) were recruited in order to investigate the features of EEG power and coherence at rest and during a working memory task. The study concluded that the MCI patients had higher EEG power at rest, and higher EEG power and coherence during working conditions. In additon, the study suggested that MCI might be associated with compensatory processes at rest and during working memory tasks.[35] Finally, using LORETA analysis, a study reported that occipital delta and alpha-1 sources in parietal, occipital, temporal, and "limbic" areas had an intermediate magnitude in MCI subjects compared to mild AD and control subjects.[36]

3.2.2.2 *Synchronization Likelihood*

SL has been investigated for MCI patients.[29,37–39] A study reported that alpha-1 SL progressively decreased across normal elderly (Nold), MCI, and mild AD subjects at the midline (Fz–Pz) and right (F4–P4) frontoparietal electrodes. The same was true for the delta SL at the right frontoparietal electrodes (F4–P4).[37] Similarly, a decreased SL has been reported in MCI patients when compared with healthy controls.[38] In another study, during working memory tasks, SL was found to be significantly higher in MCI patients compared to control subjects in the lower alpha band (8–10 Hz).[29] Finally, SL was reported to be significantly decreased in the 14–18 Hz and 18–22 Hz band in AD patients compared with both MCI subjects and healthy controls.[39] Table 3.2 has presented a summary of EEG-based FC methods for MCI.

3.2.3 Major Depressive Disorder

MDD or depression is characterized by sad mood that persists for at least two weeks. According to a survey on drug use and health, in 2015, an estimated 16.1 million adults aged 18 and older in the United States had at least one major depressive episode in the past year. This number represented 6.7% of all U.S. adults.[40]

TABLE 3.2 EEG-Based Functional Connectivity Methods for Mild Cognitive Impairment (MCI)

Year	FC Measure	Data Set Used	Finding
2006[32]	Inter- and intrahemispheric EEG coherence	MCI = 35 Healthy = 34	During working memory tasks, the inter-and intrahemispheric EEG coherences in all bands were significantly higher in the MCI group in comparison with those in the control group ($P < .05$).
2006[37]	Synchronization likelihood	MCI = 88 AD = 109 Healthy = 69	Frontoparietal coupling of brain rhythms in MCI: a multicentric EEG study.
2006[33]	Coherence	MCI at baseline = 69, at a later stage MCI = 45, MCI converted to AD = 24	Conversion from MCI to AD is predicted by sources and coherences of brain EEG rhythms.
2006[36]	Correlation	MCI = 155, mild AD = 193, and age-matched Nold subjects = 126	Sources of cortical rhythms change as a function of cognitive impairment in pathological aging: a multicenter study.
2005[38]	Global field synchronization	Healthy = 25 Patients = 183	Decreased EEG synchronization in AD and MCI patients was reported.
2005[35]	EEG power and coherence	MCI = 35 Healthy = 34	Study on EEG power and coherence in patients with MCI during working memory tasks.
2004[29]	Synchronization likelihood	MCI = 14 Healthy = 11	EEG SL in MCI and AD during working memory tasks.
2003[39]	Synchronization likelihood	MCI = 17 SCI = 20 AD = 10	SL was significantly decreased in the 14–18 Hz and 18–22 Hz bands in AD patients compared with both MCI subjects and healthy controls.
2000[34]	Coherence	MCI = 14 AD = 15 Healthy = 16	Quantitative EEG in MCI: longitudinal changes and possible prediction of AD.

3.2.3.1 *Interhemispheric Coherence*

Studies on depression have used coherence as a feature to investigate changes in FC between MDD patients and healthy controls.[41–46] More specifically, coherence-based graphs were constructed and three graph-based measures were derived: clustering coefficient (C), characteristic path length (L), and small-worldness (S).[41]

The use of LORETA has revealed that connectivity in the alpha and theta frequencies of the dorsolateral prefrontal cortex (DLPFC), dorsomedial prefrontal cortex (DMPFC), and subgenual anterior cingulate cortex (DLPFC-DMPFC-sgACC) network changed from pre- to post-treatment between (1) patients and controls and (2) responders and nonresponders.[47]

3.2.3.2 *Synchronization Likelihood*

In the context of SL,[48–50] the study[48] reported that (1) SL of male MDD subjects was significantly increased in the theta and beta frequency bands compared to the control group and (2) SL of male MDD subjects was significantly increased in the theta frequency band compared to female MDD subjects. No significant difference was found between female MDD subjects and female control group as well as between healthy male and female subjects. In a different study,[49] it was reported that in the depressed group, the mean SL was lower in the delta, theta, and sigma frequency bands. Acutely depressed patients showed a significantly lower path length in the theta and delta frequency bands, whereas the cluster coefficient showed no significant changes. The study[50] reported that major depressive patients had a significant ($P < .05$) decrease in the mean SL compared with that of healthy subjects for delta (0.5–4 Hz) band and F7, F3, Fp2, F4, and T3 for the alpha (8–13 Hz) band, which cannot be detected by conventional coherence measures.

3.2.3.3 *Partial Directed Coherence*

The application of partial directed coherence (PDC) has revealed direct connectivity.[51] Depressed patients showed lower frontal cortical interdependence in both the resting and mental arithmetic task states.

A study employing a structural synchrony approach[52] hypothesized an increase in functional connection in depressed patients. The study reported that some functional connections have shown positive correlation with the severity of depression and are thus predictive. Table 3.3 provides a summary of EEG-based FC studies for MDD.

3.2.4 Schizophrenia

Schizophrenia is a chronic and severe mental disorder that affects a person's thinking, feelings, and behavior. People with schizophrenia may seem like they have lost touch with reality. According to the National Institute of Mental Health, about 1.1% of the adult population in the United States suffer from schizophrenia.[53]

TABLE 3.3 EEG-Based Functional Connectivity (FC) Methods for Major Depressive Disorder (MDD)

Year	FC Measure	Date Set Used	Finding
2017[41]	Coherence	MDD = 37 Healthy = 37	The results of the study indicate that adding graph theoretical measures to FC does not significantly increase classification accuracy for distinguishing MDD and healthy subjects.
2017[47]	LORETA-based connectivity analysis	MDD = 447 Healthy = 336	The study emphasized that decreasing alpha connectivity could potentially serve as a treatment emergent biomarker in males only.
2016[48]	Synchronization likelihood	MDD = 37 Healthy = 37	The experimental results indicated that (1) SL of male MDD subjects was significantly increased in the theta and beta frequency bands compared to the control group and (2) SL of male MDD subjects was significantly increased in the theta frequency band compared to female MDD subjects.
2016[42]	Coherence and phase synchronization index (PSI)	MDD = 14 Healthy = 19	Patients with depression showed lower target-dependent PSI increment in the frontal-parietal/temporal/occipital electrode pairs in delta-phase synchronization than healthy participants.
2011[43]	Coherence, correlation	MDD = 108	Connectivity strengths of the right frontotemporal network at delta and theta frequencies differentiated responders and nonresponders at the eighth week of treatment: the stronger the connectivity strengths, the poorer the treatment response.
2009[49]	Synchronization likelihood	MDD = 11 Healthy = 14	In the depressed group, the mean SL was lower in the delta, theta, and sigma frequency bands. Acutely depressed patients showed a significantly lower path length in the theta and delta frequency bands, whereas the cluster coefficient showed no significant changes.
2008[51]	Partial directed coherence	MDD = 12 Healthy = 12	Depressed patients showed lower frontal cortical interdependence in both the resting and mental arithmetic task states. On the other hand, the mental arithmetic task was found to enhance inter- and intrahemispheric interactions in both groups, and such hemispheric hyperactivation is consistent with the findings from functional imaging.

(Continued)

TABLE 3.3 (Continued)

Year	FC Measure	Date Set Used	Finding
2007[50]	Synchronization likelihood	MDD = 12 Healthy = 16	Major depressive patients had a significant ($P < .05$) decrease in the mean synchronization likelihood compared with that of healthy subjects for the delta band.
2007[52]	Structural synchrony approach	MDD = 12 Healthy = 10	The study emphasized that the right anterior and left posterior brain parts may discriminate depressive patients from healthy controls.
2001[44]	Coherence	MDD = 70 Healthy = 23	Compared with controls, patients evidenced greater overall relative beta power and, at bilateral anterior regions, greater absolute beta power and faster mean total spectrum frequency. Interhemispheric alpha power asymmetry: controls exhibited relatively reduced left hemisphere activation and widespread reduced delta, theta, alpha, and beta coherence indices. Discriminant analysis correctly classified 91.3% of the patients and controls.
1995[45]	Coherence	MDD = 29Healthy = 20	Beta power was greater in patients with anxiety type depression than in normal controls.
1988[46]	Coherence	MDD = 76 Healthy = 93	A decreased interhemispheric coherence in the delta and/or theta frequency bands was present in depressed patients.

3.2.4.1 Interhemispheric Coherence and Imaginary Coherence

In schizophrenic patients, coherence and imaginary coherence were utilized for the assessment of connectivity.[54–56] The study[54] reported that schizophrenia patients showed coupling abnormalities during an auditory oddball task compared to healthy controls. In study[55], an imaginary coherence-based multivariate interaction measure (MIM) was introduced and utilized to quantify the connectivity between different brain regions. The study reported that a SZ displayed increased theta-band resting-state MIM connectivity across the midline, sensorimotor, and orbitofrontal regions as well as the left temporoparietal junction. The change was not observed in the healthy control group. Furthermore,[56] emphasized that wavelet coherence can capture the temporal profile of brain dynamics in addition to the frequency domain information.

Gamma band activation during working memory tasks implicated neuronal binding which was found abnormal during schizophrenia assessment. Similarly, in a study,[57] failure of gamma band (40 Hz rang) synchronization was reported as a marker for schizophrenia. Connectivity was quantified with phase locking and phase coherence.

3.2.4.2 *Generalized Synchronization*

Synchronization between different brain regions has been studied involving SL,[58] synchrony of brain activity,[59] and global field synchronization.[60] Abnormalities during functional disconnection have been investigated as a cause of SZ. The study[59] resulted in reduced P300 amplitude and delta and theta synchrony in patients. Furthermore, in the study,[60] a novel EEG measure was introduced—global field synchronization (GFS)—that estimates FC of brain processes in different EEG frequency bands. The study reported that the patients had significantly decreased GFS in the theta EEG frequency band, indicating a loosened FC of processes in this frequency.

Lagged phase synchronization[61] concluded that, in addition to dysfunction of the parietal regions that are part of the default mode network (DMN), resting-state disrupted connectivity of the medial temporal cortex with prefrontal areas that are either involved in the DMN or implicated in psychopathological dysfunction, it may be critical to schizophrenia-like psychosis, especially in individuals with temporal lobe epilepsy.

3.2.4.3 *Correlation and Mutual Information*

The association between different brain locations can be quantified by computing correlations between them. As a result, a correlation matrix can be formed. Further, the matirx can be converted to a weighted graph that can be utilized for the assessment of connectivity strengths between different brain locations. The study concluded that graphs in the schizophrenia group displayed lower clustering and shorter path lengths in comparison to the healthy group. MI[62] was employed to assess the FC and it was concluded that the schizophrenic patients had significantly higher interhemispheric and intrahemispheric average cross mutual information (A-CMI) values compared to the normal controls. Table 3.4 summarizes EEG-based FC measures for schizophrenia.

3.2.5 Epilepsy

Epilepsy is a metal disorder characterized by unforeseen occurrences of severe seizures. The seizures could be harmful to the patient and the surrounding environment. According to an estimate, epilepsy affects

TABLE 3.4 EEG-Based Functional Connectivity (FC) Measures for Schizophrenia

Year	FC Measure	Dataset Used	Finding
2017[54]	Coherence (MSC) and imaginary coherence (ICOH)	SZ = 28 Healthy = 51	Statistically significant modulation increases in magnitude squared coference (MSC) and ICOH for healthy controls with respect to schizophrenia (SCH) patients in the theta band and a decrease in ICOH for the beta-2 band.
2016[63]	Lagged phase synchronization[61]	SZ = 20 Healthy = 20	In patients, significantly increased current source density was found in the dominant anterior cingulate cortex. Increased connectivity between the inferior parietal lobe bilaterally and between the left inferior parietal lobe and right middle frontal gyrus was also found.
2015[55]	Imaginary coherence-based multivariate interaction measure (MIM)	SZ = 19 Healthy = 23	SZ displayed increased theta-band resting-state MIM connectivity across the midline, sensorimotor and orbitofrontal regions as well as the left temporoparietal junction. High-risk individuals (HR) displayed intermediate theta band connectivity patterns that did not differ from either SZ or healthy controls (HC). Mean theta band connectivity within the mentioned network partially mediated verbal memory deficits in SZ and HR.
2013[58]	Synchronization likelihood and graph theoretic measure	SZ = 20 Healthy = 20	Stimulus-specific decrease of path length in schizophrenia patients.
2011[61]	Lagged phase synchronization	SZ = 21 Controls = 21	The study participants included 21 patients with focal epilepsy and schizophrenia-like psychosis of epilepsy (SLPE) and in 21 clinically-matched non-psychotic epilepsy controls. The study computed source current density and functional connectivity using eLORETA software. As results, the study founded an increased theta oscillations in regions such as the medial and lateral parietal cortex in the psychotic patients relative to their nonpsychotic counterparts. Moreover, patients with psychosis showed increased beta temporo-prefrontal connectivity in the hemisphere with predominant seizure focus.

(Continued)

TABLE 3.4 (Continued)

Year	FC Measure	Dataset Used	Finding
2010[64]	Fronto-posterior coherence	SZ = 16 Healthy = 20	Event-related coherence was significantly reduced in patients during time intervals (0–250 ms poststimulus), when controls showed significant event-related coherence increases. However, patients showed significantly higher absolute coherence during the intertrial interval.
2008[59]	Synchrony of delta, theta, alpha, beta, and gamma activity in a 50 ms window around a P300 peak	SZ = 21 Healthy = 22	P300 amplitude and delta and theta synchrony were reduced in patients. Delta power and synchrony better distinguished between groups than P300 amplitude did. In healthy controls, but not patients, gamma synchrony predicted P300 amplitude. In patients, P300 and gamma synchrony are affected by independent factors.
2007[2]	Correlation matrixes converted to weighted graphs	SZ = 40 Healthy = 40	Graphs from the schizophrenia group displayed lower clustering and shorter path lengths in comparison to the healthy group.
2006[56]	Coherence	SZ = 20 Healthy = 20	The study proposed wavelet coherence analysis in combination with the graph analysis methodology. The results have shown significant task differentiation between the healthy controls and schizophernia patients. The differences were observed in gamma band atfrontal, frontal-central and temporal regions. For the above values of T (threshold), the respective L values observed were greater than those of the control subjects.
2003[57]	Phase locking and phase coherence	SZ = 14 Healthy = 12	Compared with matched control subjects, schizophrenia patients demonstrated few differences. First, the absence of the posterior component of the early visual gamma band in response to Gestalt stimuli. Second, abnormalities in the topography, latency, and frequency of the anterior component of this response. Third, delayed onset of phase coherence changes. Forth, the pattern of anterior-posterior coherence increases in response to Gestalt stimuli found in the controls was replaced by a pattern of interhemispheric coherence decreases in patients.

(Continued)

TABLE 3.4 (Continued)

Year	FC Measure	Dataset Used	Finding
2002[62]	Mutual information	SZ = 10 Healthy = 10	In the T5 and C3 electrodes, the schizophrenic patients had lower complexity compared to normal controls. The schizophrenic patients had significantly higher interhemispheric and intrahemispheric A-CMI values than the normal controls.
2001[60]	Global field synchronization	SZ = 11 Healthy = 19	In comparison to age- and sex-matched controls, patients had significantly decreased GFS in the theta EEG frequency band, indicating a loosened FC of processes in this frequency.

more than 500,000 people in the United Kingdom. This means that almost one in every 100 people has the condition.[65]

3.2.5.1 Interhemispheric Coherence

Coherence-based studies for epilepsy have reported abnormal brain connectivity during epilepsy.[66,67] For example, in study,[66] seven different measures of quantifying synchronous oscillatory activity were computed. These measures include coherence, a coding-based measure known as minimum description length (MDL), and the Geweke alternative[68]. Additionally, a robust phase coupling measure known as phase locking value (PLV), a cortical synchrony measure defined from the embedding dimension in state-space called S-estimator was also computed. Moreover, a reliable way of assessing generalized synchronization also in state-space and an unbiased alternative is SL. From the results on the actual data, higher frequency band gamma2 was mostly apparent in occipital parietal lobes during fractal tests. In a study,[67] mean phase coherence was used as a statistical measure for phase synchronization. The study reported distinct differences in the degree of synchronization between recordings from seizure-free intervals and those before an impending seizure, indicating an altered state of brain dynamics prior to seizure activity. In a different study,[69] correlation was utilized as a measure to quantify the FC. As a result, the study reported enhanced FC during epilepsy.

3.2.5.2 Synchronization Likelihood

Studies involving SL[70,71] consisted of the construction of networks based on SL matrices. The study[71] reported that the neuronal network changed during seizure activity, with an increase of clustering coefficient (C) and shortest path length (L) most prominent in the alpha, theta, and delta frequency bands during and after the seizure. A similar study[70] reported that during the absence of seizures there was an increase of synchronization in all frequency bands, seen most clearly in SL-based networks, and the functional network topology was changed to a more ordered pattern, with an increase of C/C-s and L/L-s. The mean (C-s and L-s) of these networks were used as a reference value for C and L (C/C-s and L/L-s). These normalized values were used during the study.

Different topological network metrics such as minimum spanning tress (MST) and betweenness centrality were computed. The study[72] reported that network alterations between groups were only identified by MST metrics and were most pronounced in the delta band, in which a loss of network integration and a significant lower betweenness centrality was found in children with focal epilepsies compared to healthy controls ($P < .01$).

3.2.5.3 *Correlation*

The correlation between LORETA sources has been evaluated in different studies.[73,74] In a study,[73] increased EEG FC was reported while evaluating the correlation between different LORETA-based current source time series. In study,[74] remote EEG synchronization (intrahemispheric, cortico-cortical EEG FC, EEGfC) was computed using the LSC (LORETA source correlation) method. The study resulted in individual results being presented. Abnormal but topographically dissimilar LORETA and LSC findings were found at the onset of the disease. The study concluded that EEG-based local and remote connectivity (EEGfC) are appropriate tools to describe network dynamics in epilepsy. In a study,[75] the detection of epilepsy was performed based on FC features and resulted in a 92.8% accuracy of classification between the epileptic group and the healthy controls (Table 3.5)

3.2.6 Alcoholism

According to the 2015 National Survey on Drug Use and Health (NSDUH), 15.1 million adults aged 18 and older (6.2% of this age group) had alcohol use disorder (AUD) and an estimated 623,000 adolescents between the ages of 12 and 17 (2.5% of this age group) had AUD.[77] A comprehensive review has been provided.[78]

3.2.6.1 *Interhemispheric Coherence*

Coherence has been used for alcoholism.[79] The study hypothesized that acute alcohol intake would increase the FC of the human brain

TABLE 3.5 EEG-Based Functional Connectivity (FC) Measures for Epilepsy

Year	FC Measure	Dataset Used	Finding
2016[71]	Conventional network metrics and minimum spanning tree (MST) metrics were computed	Epileptic = 89 Healthy = 179	A loss of network integration and a significant lower betweenness centrality was found in children with focal epilepsies compared to healthy controls ($P < .01$). A reversed group difference was found in the upper alpha band.
2016[72]	Correlation of (LORETA) defined current source density time series were computed between two cortical areas	Epileptic = 17 Healthy = 19	(1) Bilaterally increased beta EEGfC occurred in the epileptic subjects as compared to the controls. (2) Locally increased EEGfC emerged in all frequency bands in the right parietal area.
2015[69]	Coherence and synchronization likelihood	Children with the absence of seizures = 11	The network became more regularized in weighted and unweighted analyses when compared to a more randomized pre-ictal network configuration.
2015[74]	This study used probability of recurrence and the correlation between electrodes	Epileptic = 14	Performance results show an accuracy of 92.8% with a sensitivity of 85.7% and a specificity of 100% when tested on 14 subjects.
2013[73]	LORETA and LORETA source correlation	Children with benign epilepsy = 4	Individual results were presented. Abnormal but topographically dissimilar LORETA and LSC findings were found at the onset of the disease. The disappearance of the initial abnormalities was found in Setting No. 2. An unforeseen finding was the presence of abnormal EEGfC results in Setting No. 2.
2008[68]	Correlation mesial temporal lobe epilepsy (MTLE), nonmesial temporal lobe epilepsy (NMTLE)	Epileptic = 21	Power spectral density analysis showed a significant decrease in the theta frequency sub-band ($P = .01$) in the MTLE group. Nonlinear correlation (h^2) values were found to be higher in the MTLE group than in the NMTLE group ($P = .0014$). This effect was significant for theta, alpha, beta, and gamma frequencies.

(Continued)

TABLE 3.5 (Continued)

Year	FC Measure	Dataset Used	Finding
2007[70]	Synchronization likelihood Clustering coefficient (C) Shortest path length (L)	Epileptic = 7	The neuronal network changed during seizure activity, with an increase of C and L most prominent in the alpha, theta, and delta frequency bands during and after seizure.
2007[75]	Mean phase coherence algorithm	Epileptic = 9	This analysis revealed areas of elevated local synchrony or "hypersynchrony," which had persistent spatiotemporal characteristics that were unique to each patient.
2006[66]	Coherence, MDL, and PLV S-estimator, synchronization likelihood	Epileptic = 20 Healthy = 20	The results on the actual data suggest higher frequency band gamma-2 was mostly apparent in occipital parietal lobes during fractal tests.
2000[67]	Mean phase coherence	Epileptic = 17	It was concluded that phase synchronization between two EEG channels as measured by the mean phase coherence R appears to be related to pathological findings in epilepsy patients.

resting-state network (RSN). The authors calculated coherence using standardized low resolution electromagnetic tomography (sLORETA) solutions to form cortical networks over several frequency bands, which were then analyzed in the context of FC and graph theory. Graph parameters were accordingly altered in these bands, quantifying the effect of alcohol on the structure of brain networks; global efficiency and density were higher and path length was lower under alcoholism (versus placebo, $P < .05$).

3.2.6.2 *Synchronization Likelihood*

In general, alcoholics have exhibited decreased or abnormal synchronization during rest conditions. Many research studies have used SL as a measure for quantification of FC in alcoholics.[80–83] Moreover, other studies have utilized phase synchrony[84] and synchronized EEG patterns[85] for the assessment of the differences in normal and alcoholic brains. Table 3.6 summarizes EEG-based FC measures during alcoholism.

TABLE 3.6 EEG-Based Functional Connectivity (FC) Measures for Alcoholism

Year	FC Measure	Dataset Used	Finding
2017[79]	Synchronization likelihood	Alcoholics = 28 Healthy = 28	Synchronization for the alcoholic group was lower in multiple bands than in the control group when performing the same cognitive task.
2016[80]	Synchronization likelihood	Alcoholics = 15 Healthy = 15	Increased absolute and relative beta power in AUD patients compared to matched controls, and reduced FC in AUD patients predominantly in the beta and alpha bands.
2012[78]	Magnitude square coherence (MSC) on sLORETA solutions	Healthy subjects (light drinkers) = 26	MSC was increased ($P < .05$, corrected with false discovery rate [FDR] corrected) in alpha, beta (eyes open), and theta bands (eyes closed) following acute alcohol intake. Graph parameters were accordingly altered in these bands quantifying the effect of alcohol on the structure of brain networks; global efficiency and density were higher and path length was lower under alcoholism (versus placebo, $P < .05$).
2012[83]	Phase synchrony	Alcoholics = 77 Healthy = 45	Phase synchrony computed for 2.34 s-long overlapping EEG fragments was lower for alcoholics than for controls when evaluated in alpha-2 and beta-1 rhythms and for specific electrode pairs.
2006[81]	Synchronization likelihood		Both male and female heavy drinkers displayed a loss of lateralization in alpha (8–12 Hz) and slow-beta (12–20 Hz) synchronization. In addition, moderately and heavily drinking males had lower fast-beta (20–30 Hz) synchronization than lightly drinking males.

(Continued)

TABLE 3.6 (Continued)

Year	FC Measure	Dataset Used	Finding
2004[82]	Synchronization likelihood	Light drinkers = 11 Heavy drinkers = 11	Heavily drinking students had more synchronization in the theta (4–8 Hz) and gamma (30–45 Hz) bands than lightly drinking students during eyes closed state, both with and without a mental-rehearsal task.
1981[84]	Synchronized EEG pattern	Alcoholics = 115 (78 males and 37 females) and matched controls	Male alcoholics did not differ from their controls, female patients showed a shift from the alpha and theta to the beta bands of the brain wave pattern.

3.3 OPEN-SOURCE TOOLBOXES

Recently, there has been a surge of toolboxes that include indexes of brain connectivity and made publicly available as well as published in literature.[86–89] However, most of them either focus on a special type of connectivity index (e.g., linear indexes[90]) and/or only include a subset of indexes as part of a more general purpose toolbox whose main aim is, say, the analysis of EEG and/or MEG.[86,88,91,92] The availability of multiple toolboxes allows researchers to perform quick analysis of data. Usually, the choice of method involves the underlying hypothesis to be tested. On the other hand, the choice of toolbox can significantly reduce data analysis time. The aim of this section is to review open-source MATLAB-based toolboxes that can analyze EEG and ERP data. In addition, this section details the most cited toolboxes as well as their limitations.

3.3.1 Extended Multivariate Autoregressive Toolbox

The extended multivariate autoregressive (eMVAR) toolbox is a freely available Matlab-based toolbox. in addition, it is distributed under a GNU general public user license.[93] The toolbox is based on an extended multivariate autoregressive concept which is an extension of the MVAR model. More specifically, the eMVAR model considers the effects of instantaneous causality in the computation of the causality measure. Furthermore, the toolbox incorporates methods such as extended PDC (ePDC), extended DC (eDC), normalized lagged directed coherence (nDC), and normalized lagged PDC (nPDC). The toolbox only provides frequency domain causality measures. The online help is sufficient for using the toolbox for customized datasets; however, the graphical user interface (GUI) is not developed.

3.3.2 The Granger Causality Connectivity Analysis Toolbox

The Granger causality connectivity analysis (GCCA) toolbox is a MATLAB-based toolbox and freely available and distributed under a GNU general public user license.[90] The toolbox provides the option to analyze EEG, ERP, MEG, and fMRI datasets. On the contrary, the toolbox mainly focuses on the computation of G-causality from data. The core functions implement G-causality analysis given multivariate time series data. Other functions test whether the provided data satisfy necessary assumptions, assess the statistical significance and validity of inferred interactions, generate network-level descriptions of patterns of causal interactions, and graphically display analysis results. The help manual is sufficient for using the toolbox for customized datasets; however, the GUI is not developed.

3.3.3 The Multivariate Granger Causality Toolbox

The multivariate Granger causality toolbox (MVGC) is a MATLAB-based toolbox and freely available and distributed under a GNU general public user license.[94] The MVGC toolbox is an extended version of the GCCA toolbox. The MVGC MATLAB toolbox implements numerical routines for calculating MVGC from time series data, both unconditional and conditional, in time and frequency domains. The help manual is sufficient for using the toolbox for customized datasets; however, the GUI is not developed.

3.3.4 HERMES Toolbox

HERMES is a MATLAB-based toolbox and freely available and distributed under a GNU general public user license that supports multiple time series analysis.[95] The HERMES toolbox offers multiple methods for the assessment of brain function and EC. For example, it includes the conventional methods such as correlation, coherence, PLI and Granger causality methods such as directed coherence, PDC, directed transfer function, partial directed transfer function. The toolbox provides a GUI. In addition, it allows the user to provide input EEG data in the form of pdf files in the form of groups. Group comparisons can be performed for a single participant or a group of participants. Finally, the results are

TABLE 3.7 MATLAB-Based Open Source Toolboxes for Quantification of Brain Connectivity

Reference Publication	Methods Included in the Toolbox	Download Address	Limitations
eMVAR[92]	Extended PDC (ePDC), extended DC (eDC), normalized lagged directed coherence (nDC), normalized lagged PDC (nPDC)	http://www.science.unitn.it/biophysicslab/research/sigpro/eMVAR.html	Relatively smaller in size than the HERMES toolbox. No GUI is provided.
GCCA toolbox[89]	The core functions implement G-causality analysis given multivariate time series data	http://www.anilseth.com	Relatively smaller in size than the HERMES toolbox. No GUI is provided.
MVGC toolbox[93]	The MVGC toolbox is an extended version of the GCCA toolbox	http://www.sussex.ac.uk/sackler/	Relatively smaller in size than the HERMES toolbox. No GUI is provided.
HERMES toolbox[94]	Conventional methods (correlation, coherence, PLI), Granger causality methods (directed coherence, partial directed coherence, directed transfer function, partial directed transfer function)	http://hermes.ctb.upm.es/downloads/	Although the toolbox is rich in the presentation of various methods, the outcome of the toolbox is available in the form of figures only; the output data cannot be exported as numbers.
eConnectome[95]	Directed transfer function and adaptive directed transfer function	http://econnectome.umn.edu/	Relatively smaller than the HERMES toolbox.
BSMART[96]	Time series data importing/exporting, preprocessing (normalization and trend removal), autoregressive modeling (multivariate/bivariate model estimation and validation), spectral quantity estimation (auto power, coherence, and Granger causality spectra), and network analysis (including coherence and causality networks)	http://www.brain-smart.org/	Relatively smaller than the HERMES toolbox.

provided as topographic plots that show brain network connections between electrodes (defined in the sensor location file). The generated scalp location plots can be saved as pdf files and can be presented as results.

3.3.5 The eConnectome Toolbox

The eConnectome toolbox is a MATLAB-based toolbox and freely available and distributed under a GNU general public user license.[96] The eConnectome toolbox is a MATLAB-based toolbox that uses EEG and electrocorticography (ECoG) data as input data. It computes brain connectivity (EC) using methods such as directed transfer function and adaptive directed transfer function. Other methods include scalp spatial mapping, cortical source estimation.

3.3.6 The BSMART Toolbox

The brain system for multivariate auto regressive time series (BSMART) toolbox is a MATLAB-based and C toolbox and freely available and distributed under a GNU general public user license.[97] The toolbox includes methods such as time series data importing/exporting, preprocessing (normalization and trend removal), auto regressive (AR) modeling (multivariate/bivariate model estimation and validation), spectral quantity estimation (auto power, coherence, and Granger causality spectra), network analysis (including coherence and causality networks), and visualization (including data, power, coherence, and causality views). Table 3.7 provides a summary of these toolboxes.

3.4 SUMMARY

In this chapter, clinical applications of EEG-based FC methods are discussed in the context of common mental illnesses such as bipolar disorder, epilepsy, MDD, schizophrenia, and Alzheimer. The purpose of this description is to acquaint readers with the significance of different FC measures during the diagnosis process. In addition, the most commonly used and downloaded open-source Matlab toolboxes are discussed. Finally, the main challenges faced by different measures are discussed that can pose a challenge during the interpretation of results and could lead to false brain connectivity analysis. In short, the chapter combined various aspects of brain connectivity that should be known to new researchers aiming to working in this area.

References

1. Bastos AM, Schoffelen J-M. A tutorial review of functional connectivity analysis methods and their interpretational pitfalls. *Front syst neurosci.* 2015;9.
2. Rubinov M, Knock SA, Stam CJ, et al. Small-world properties of nonlinear brain activity in schizophrenia. *Hum Brain Mapp.* 2009;30:403–416.
3. Friston KJ. Functional and effective connectivity: a review. *Brain Connect.* 2011;1:13–36.
4. Friston KJ, Harrison L, Penny W. Dynamic causal modelling. *Neuroimage.* 2003;19:1273–1302.
5. Rogers BP, Morgan VL, Newton AT, Gore JC. Assessing functional connectivity in the human brain by fMRI. *Magnet Resonl Imaging.* 2007;25:1347–1357.
6. Rangaswamy M, Porjesz B, Chorlian DB, et al. Theta power in the EEG of alcoholics. *Alcohol Clin Exp Res.* 2003;27:607–615.
7. Mumtaz W, Malik AS, Yasin MAM, Xia L. Review on EEG and ERP predictive biomarkers for major depressive disorder. *Biomed Signal Proces Control.* 2015;22:85–98.
8. Lenox RH, Gould TD, Manji HK. Endophenotypes in bipolar disorder. *Am J Med Genet A.* 2002;114:391–406.
9. Sakkalis V. Review of advanced techniques for the estimation of brain connectivity measured with EEG/MEG. *Comput Biol Med.* 2011;41:1110–1117.
10. Blinowska KJ. Review of the methods of determination of directed connectivity from multichannel data. *Med Biol Eng Comput.* 2011;49:521–529.
11. Bressler SL, Seth AK. Wiener–Granger causality: a well established methodology. *Neuroimage.* 2011;58:323–329.
12. Zeng L-L, Shen H, Liu L, et al. Identifying major depression using whole-brain functional connectivity: a multivariate pattern analysis. *Brain.* 2012;135:1498–1507.
13. Dauwels J, Vialatte F, Musha T, Cichocki A. A comparative study of synchrony measures for the early diagnosis of Alzheimer's disease based on EEG. *NeuroImage.* 2010;49:668–693.
14. Boly M, Moran R, Murphy M, et al. Connectivity changes underlying spectral EEG changes during propofol-induced loss of consciousness. *J Neurosci.* 2012;32:7082–7090.
15. A. s. Association. (3 June). *What is Alzheimer?* Available: http://www.alz.org/alzheimers_disease_what_is_alzheimers.asp.
16. Wang R, Wang J, Yu H, Wei X, Yang C, Deng B. Decreased coherence and functional connectivity of electroencephalograph in Alzheimer's disease. *Chaos.* 2014;24:033136.
17. Güntekin B, Saatçi E, Yener G. Decrease of evoked delta, theta and alpha coherences in Alzheimer patients during a visual oddball paradigm. *Brain Res.* 2008;1235:109–116.
18. Chan H-L, Chu J-H, Fung H-C, et al. Brain connectivity of patients with Alzheimer's disease by coherence and cross mutual information of electroencephalograms during photic stimulation. *Med Eng Phys.* 2013;35:241–252.
19. Jelles B, Scheltens P, Van der Flier W, Jonkman E, da Silva FL, Stam C. Global dynamical analysis of the EEG in Alzheimer's disease: frequency-specific changes of functional interactions. *Clin Neurophysiol.* 2008;119:837–841.
20. Hidasi Z, Czigler B, Salacz P, Csibri É, Molnár M. Changes of EEG spectra and coherence following performance in a cognitive task in Alzheimer's disease. *Int J Psychophysiol.* 2007;65:252–260.
21. Zheng-Yan J. Abnormal cortical functional connections in Alzheimer's disease: analysis of inter-and intra-hemispheric EEG coherence. *J Zhejiang Univ Sci B.* 2005;6:259–264.
22. Hogan MJ, Swanwick GR, Kaiser J, Rowan M, Lawlor B. Memory-related EEG power and coherence reductions in mild Alzheimer's disease. *Int J Psychophysiol.* 2003;49:147–163.
23. Kwak YT, Hahm DS, Han IW, Sam BJ. EEG coherence in Alzheimer's disease. *J Korean Neurol Assoc.* 2003;21:163–168.

24. Anghinah R, Kanda PAM, Jorge MS, Lima EEPD, Pascuzzi L, Melo ACDP. Alpha band coherence analysis of EEG in healthy adult and Alzheimer's type dementia subjects. *Arq Neuropsiquiatr*. 2000;58:272–275.
25. Başar E, Femir B, Emek-Savaş DD, Güntekin B, Yener GG. Increased long distance event-related gamma band connectivity in Alzheimer's disease. *NeuroImage Clin*. 2017;14:580–590.
26. Yu M, Gouw AA, Hillebrand A, et al. Different functional connectivity and network topology in behavioral variant of frontotemporal dementia and Alzheimer's disease: an EEG study. *Neurobiol Aging*. 2016;42:150–162.
27. Engels MM, Stam CJ, van der Flier WM, Scheltens P, de Waal H, van Straaten EC. Declining functional connectivity and changing hub locations in Alzheimer's disease: an EEG study. *BMC Neurol*. 2015;15:145.
28. Stam C, Montez T, Jones B, et al. Disturbed fluctuations of resting state EEG synchronization in Alzheimer's disease. *Clin Neurophysiol*. 2005;116:708–715.
29. Pijnenburg Y, Vd Made Y, Van Walsum AVC, Knol D, Scheltens P, Stam C. EEG synchronization likelihood in mild cognitive impairment and Alzheimer's disease during a working memory task. *Clin Neurophysiol*. 2004;115:1332–1339.
30. Jeong J, Gore JC, Peterson BS. Mutual information analysis of the EEG in patients with Alzheimer's disease. *Clin Neurophysiol*. 2001;112:827–835.
31. A. s. Society. (3 June). *What is mild cognitive impairment?* Available: https://www.alzheimers.org.uk/info/20007/types_of_dementia/16/mild_cognitive_impairment_mci.
32. Jiang Z-Y, Zheng L-L. Inter-and intra-hemispheric EEG coherence in patients with mild cognitive impairment at rest and during working memory task. *J Zhejiang Univ Sci B*. 2006;7:357–364.
33. Rossini P, Del Percio C, Pasqualetti P, et al. Conversion from mild cognitive impairment to Alzheimer's disease is predicted by sources and coherence of brain electroencephalography rhythms. *Neuroscience*. 2006;143:793–803.
34. Jelic V, Johansson S, Almkvist O, et al. Quantitative electroencephalography in mild cognitive impairment: longitudinal changes and possible prediction of Alzheimer's disease. *Neurobiol Aging*. 2000;21:533–540.
35. Jiang Z-Y. Study on EEG power and coherence in patients with mild cognitive impairment during working memory task. *J Zhejiang Univ Sci B*. 2005;6:1213–1219.
36. Babiloni C, Binetti G, Cassetta E, et al. Sources of cortical rhythms change as a function of cognitive impairment in pathological aging: a multicenter study. *Clin Neurophysiol*. 2006;117:252–268.
37. Babiloni C, Ferri R, Binetti G, et al. Fronto-parietal coupling of brain rhythms in mild cognitive impairment: a multicentric EEG study. *Brain Res Bull*. 2006;69:63–73.
38. König T, Prichep L, Dierks T, et al. Decreased EEG synchronization in Alzheimer's disease and mild cognitive impairment. *Neurobiol Aging*. 2005;26:165–171.
39. Stam C, Van Der Made Y, Pijnenburg Y, Scheltens P. EEG synchronization in mild cognitive impairment and Alzheimer's disease. *Acta Neurol Scand*. 2003;108:90–96.
40. DBSA. (3 june). *Depression Statistics*. Available: http://www.dbsalliance.org/site/PageServer?pagename = education_statistics_depression.
41. Orgo L, Bachmann M, Kalev K, Järvelaid M, Raik J, Hinrikus H, "Resting EEG functional connectivity and graph theoretical measures for discrimination of depression." in Biomedical & Health Informatics (BHI), 2017 IEEE EMBS International Conference on. Orlando, FL, USA; 2017: 389–392.
42. Li Y, Kang C, Qu X, Zhou Y, Wang W, Hu Y. Depression-related brain connectivity analyzed by EEG event-related phase synchrony measure. *Front Hum Neurosci*. 2016;10.
43. Lee T-W, Wu Y-T, Yu YW-Y, Chen M-C, Chen T-J. The implication of functional connectivity strength in predicting treatment response of major depressive disorder: a resting EEG study. *Psychiat Res-Neuroim*. 2011;194:372–377.

44. Knott V, Mahoney C, Kennedy S, Evans K. EEG power, frequency, asymmetry and coherence in male depression. *Psychiat Res-Neuroim*. 2001;106:123–140.
45. Yamada M, Kimura M, Mori T, Endo S. EEG power and coherence in presenile and senile depression. Characteristic findings related to differences between anxiety type and retardation type. *Nihon Ika Daigaku Zasshi*. 1995;62:176–185.
46. Lieber AL. Diagnosis and subtyping of depressive disorders by quantitative electroencephalography: II. Interhemispheric measures are abnormal in major depressives and frequency analysis may discriminate certain subtypes. *Hillside J Clin Psychiat*. 1988.
47. Iseger TA, Korgaonkar MS, Kenemans JL, et al. EEG connectivity between the subgenual anterior cingulate and prefrontal cortices in response to antidepressant medication. *Eur Neuropsychopharmacol*. 2017;27:301–312.
48. Orgo L, Bachmann M, Kalev K, Hinrikus H, Järvelaid M, "Brain functional connectivity in depression: Gender differences in EEG." in *Biomedical Engineering and Sciences (IECBES), 2016 IEEE EMBS Conference on*. Kuala Lumpur, Malaysia; 2016: 270–273.
49. Leistedt SJ, Coumans N, Dumont M, Lanquart JP, Stam CJ, Linkowski P. Altered sleep brain functional connectivity in acutely depressed patients. *Hum Brain Mapp*. 2009;30:2207–2219.
50. Park C-A, Kwon R-J, Kim S, et al., "Decreased phase synchronization of the EEG in patients with major depressive disorder," In: *World Congress on Medical Physics and Biomedical Engineering 2006*, 2007, Springer, Berlin, 1095–1098.
51. Sun Y, Li Y, Zhu Y, Chen X, Tong S. Electroencephalographic differences between depressed and control subjects: an aspect of interdependence analysis. *Brain Res Bull*. 2008;76:559–564.
52. Fingelkurts AA, Fingelkurts AA, Rytsälä H, Suominen K, Isometsä E, Kähkönen S. Impaired functional connectivity at EEG alpha and theta frequency bands in major depression. *Hum Brain Mapp*. 2007;28:247–261.
53. NIMH. (2016, 3 June). Schizophrenia. Available: https://www.nimh.nih.gov/health/topics/schizophrenia/index.shtml.
54. Núñez P, Poza J, Bachiller A, et al. *Analysis of Functional Connectivity During an Auditory Oddball Task in Schizophrenia*. Converging Clinical and Engineering Research on Neurorehabilitation. II, ed Springer; 2017:751–755.
55. Andreou C, Leicht G, Nolte G, et al. Resting-state theta-band connectivity and verbal memory in schizophrenia and in the high-risk state. *Schizophr Res*. 2015;161:299–307.
56. Sakkalis V, Oikonomou T, Pachou E, Tollis I, Micheloyannis S, Zervakis M, "Time-significant wavelet coherence for the evaluation of schizophrenic brain activity using a graph theory approach." In: *Engineering in Medicine and Biology Society, 2006. EMBS'06. 28th Annual International Conference of the IEEE*. New York, NY, USA; 2006: 4265–4268.
57. Spencer KM, Nestor PG, Niznikiewicz MA, Salisbury DF, Shenton ME, McCarley RW. Abnormal neural synchrony in schizophrenia. *J Neurosci*. 2003;23:7407–7411.
58. Fogelson N, Li L, Li Y, Fernandez-del-Olmo M, Santos-Garcia D, Peled A. Functional connectivity abnormalities during contextual processing in schizophrenia and in Parkinson's disease. *Brain Cognition*. 2013;82:243–253.
59. Ford JM, Roach BJ, Hoffman RS, Mathalon DH. The dependence of P300 amplitude on gamma synchrony breaks down in schizophrenia. *Brain Res*. 2008;1235:133–142.
60. König T, Lehmann D, Saito N, Kuginuki T, Kinoshita T, Koukkou M. Decreased functional connectivity of EEG theta-frequency activity in first-episode, neuroleptic-naıve patients with schizophrenia: preliminary results. *Schizophr Res*. 2001;50:55–60.
61. Canuet L, Ishii R, Pascual-Marqui RD, et al. Resting-state EEG source localization and functional connectivity in schizophrenia-like psychosis of epilepsy. *PLoS ONE*. 2011;6:e27863.
62. Na SH, Jin S-H, Kim SY, Ham B-J. EEG in schizophrenic patients: mutual information analysis. *Clin Neurophysiol*. 2002;113:1954–1960.

63. Umesh DS, Tikka SK, Goyal N, Nizamie SH, Sinha VK. Resting state theta band source distribution and functional connectivity in remitted schizophrenia. *Neurosci Lett*. 2016;630:199–202.
64. Sharma A, Weisbrod M, Kaiser S, Markela-Lerenc J, Bender S. Deficits in fronto-posterior interactions point to inefficient resource allocation in schizophrenia. *Acta Psychiatr Scand*. 2011;123:125–135.
65. NHS. (2014, 3 June). *Epilepsy* Available: http://www.nhs.uk/conditions/Epilepsy/Pages/Introduction.aspx.
66. Sakkalis V, Giurcăneacu C, Xanthopoulos P, et al. Assessment of linear and non-linear EEG synchronization measures for evaluating mild epileptic signal patterns. *Proc ITAB*. 2006;26–28.
67. Mormann F, Lehnertz K, David P, Elger CE. Mean phase coherence as a measure for phase synchronization and its application to the EEG of epilepsy patients. *Physica D: Nonlinear Phenomena*. 2000;144:358–369.
68. Geweke J. Measurement of linear dependence and feedback between multiple time series. *Journal of the American Statistical Association*. 1982;77(378):304–313.
69. Bettus G, Wendling F, Guye M, et al. Enhanced EEG functional connectivity in mesial temporal lobe epilepsy. *Epilepsy Res*. 2008;81:58–68.
70. Ponten S, Douw L, Bartolomei F, Reijneveld J, Stam C. Indications for network regularization during absence seizures: weighted and unweighted graph theoretical analyses. *Exp Neurol*. 2009;217:197–204.
71. Ponten S, Bartolomei F, Stam C. Small-world networks and epilepsy: graph theoretical analysis of intracerebrally recorded mesial temporal lobe seizures. *Clin Neurophysiol*. 2007;118:918–927.
72. van Diessen E, Otte WM, Stam CJ, Braun KP, Jansen FE. Electroencephalography based functional networks in newly diagnosed childhood epilepsies. *Clin Neurophysiol*. 2016;127:2325–2332.
73. Clemens B, Puskás S, Spisák T, et al. Increased resting-state EEG functional connectivity in benign childhood epilepsy with centro-temporal spikes. *Seizure*. 2016;35:50–55.
74. Clemens B, Puskás S, Besenyei M, Spisák T, Emri M, Fekete I. Remission of benign epilepsy with rolandic spikes: an EEG-based connectivity study at the onset of the disease and at remission. *Epilepsy Res*. 2013;106:128–135.
75. Rajaei H, Cabrerizo M, Sargolzaei S, Pinzon-Ardila A, Gonzalez-Arias S, Adjouadi M, "Pediatric epilepsy: clustering by functional connectivity using phase synchronization." In: *Biomedical Circuits and Systems Conference (BioCAS), 2015 IEEE*. Atlanta, GA, USA: 1–4.
76. Schevon C, Cappell J, Emerson R, et al. Cortical abnormalities in epilepsy revealed by local EEG synchrony. *Neuroimage*. 2007;35:140–148.
77. NIAAA. (2017, 3 June). *Alcohol Facts and Statisitics*. Available: https://www.niaaa.nih.gov/alcohol-health/overview-alcohol-consumption/alcohol-facts-and-statistics.
78. Porjesz B, Begleiter H. Alcoholism and human electrophysiology. *Alcohol Res Health*. 2003;27:153–160.
79. Lithari C, Klados MA, Pappas C, et al. Alcohol affects the brain's resting-state network in social drinkers. *PLoS ONE*. 2012;7:e48641.
80. Cao R, Deng H, Wu Z, Liu G, Guo H, Xiang J. Decreased Synchronization in Alcoholics Using EEG. *IRBM*. 2017;38:63–70.
81. Herrera-Díaz A, Mendoza-Quiñones R, Melie-Garcia L, et al. Functional connectivity and quantitative EEG in women with alcohol use disorders: a resting-state study. *Brain Topogr*. 2016;29:368–381.
82. de Bruin EA, Stam CJ, Bijl S, Verbaten MN, Kenemans JL. Moderate-to-heavy alcohol intake is associated with differences in synchronization of brain activity during rest and mental rehearsal. *Int J Psychophysiol*. 2006;60:304–314.

83. de Bruin EA, Bijl S, Stam CJ, Böcker KB, Kenemans JL, Verbaten MN. Abnormal EEG synchronisation in heavily drinking students. *Clin Neurophysiol*. 2004;115:2048–2055.
84. Tcheslavski GV, Gonen FF. Alcoholism-related alterations in spectrum, coherence, and phase synchrony of topical electroencephalogram. *Comp Biol Med*. 2012;42:394–401.
85. Propping P, Krüger J, Mark N. Genetic disposition to alcoholism. An EEG study in alcoholics and their relatives. *Hum Genet*. 1981;59:51–59.
86. Oostenveld R, Fries P, Maris E, Schoffelen J-M. FieldTrip: open source software for advanced analysis of MEG, EEG, and invasive electrophysiological data. *Comput Intel Neurosci*. 2011;2011:1.
87. Rose J, Otto T, Dittrich L. The Biopsychology-Toolbox: a free, open-source Matlab-toolbox for the control of behavioral experiments. *J Neurosci Methods*. 2008;175:104–107.
88. Tadel F, Baillet S, Mosher JC, Pantazis D, Leahy RM. Brainstorm: a user-friendly application for MEG/EEG analysis. *Comput Intel Neurosci*. 2011;2011:8.
89. Zhou D, Thompson WK, Siegle G. MATLAB toolbox for functional connectivity. *Neuroimage*. 2009;47:1590–1607.
90. Seth AK. A MATLAB toolbox for Granger causal connectivity analysis. *J Neurosci Methods*. 2010;186:262–273.
91. Delorme A, Makeig S. EEGLAB: an open source toolbox for analysis of single-trial EEG dynamics including independent component analysis. *J Neurosci Methods*. 3/15/2004;134:9–21.
92. Delorme A, Mullen T, Kothe C, et al. EEGLAB, SIFT, NFT, BCILAB, and ERICA: new tools for advanced EEG processing. *Comput Intel Neurosci*. 2011;2011:10.
93. Faes L, Erla S, Porta A, Nollo G. A framework for assessing frequency domain causality in physiological time series with instantaneous effects. *Philos Trans R Soc Lond A*. 2013;371:20110618.
94. Barnett L, Seth AK. The MVGC multivariate Granger causality toolbox: a new approach to Granger-causal inference. *J Neurosci Methods*. 2014;223:50–68.
95. Niso G, Bruña R, Pereda E, et al. HERMES: towards an integrated toolbox to characterize functional and effective brain connectivity. *Neuroinformatics*. 2013;11:405–434.
96. He B, Dai Y, Astolfi L, Babiloni F, Yuan H, Yang L. eConnectome: a MATLAB toolbox for mapping and imaging of brain functional connectivity. *J Neurosci Methods*. 2011;195:261–269.
97. Cui J, Xu L, Bressler SL, Ding M, Liang H. BSMART: a Matlab/C toolbox for analysis of multichannel neural time series. *Neural Netw*. 2008;21:1094–1104.

CHAPTER

4

Pathophysiology of Depression

4.1 INTRODUCTION

This chapter elaborates on various underlying factors that could become causes or effects of depression. Specifically, depressed patients have exhibited brain volume reductions. The literature has implicated brain regions such as the basal ganglia, frontal cortex, amygdala, and hippocampus. In addition, brain volume loss could be due to brain cell death. Details on brain volume abnormalities are presented in Section 4.2. The literature also shows various factors other than brain volume loss that include: (1) genetic and nongenetic factors; (2) effects of stress hormones on depression; (3) effects of different monoamines; (4) dysfunction in specific brain regions; (5) neurotrophic hypothesis; (6) lessened GABAergic activity; (7) dysregulations of the glutamate system; and/or (8) damaged circadian rhythms. Details of each of these factors are provided in Section 4.3.

4.2 BRAIN VOLUME ABNORMALITIES DURING DEPRESSION

In this section, major depressive disorder (MDD)-related research studies are discussed that have reported a loss of volume in different brain regions. During MDD, structural changes were mainly related to volume loss in the Limbic-Cortical-Striatal-Pallidal-Thalamic (LCSPT) tract. The tract consists of two arms. The first arm involves the limbic-thalamic-cortical branch composed of amygdala and hippocampus, mediodorsal nucleus of the thalamus, and medial and ventrolateral prefrontal cortex, while the second arm includes the limbic-striatal-pallidal thalamic branch.

However, even in healthy persons, small structural changes did not always indicate depression. Hence, a direct correlation between depression and structural impairments in LCSPT structures were not found.

EEG-Based Experiment Design for Major Depressive Disorder
DOI: https://doi.org/10.1016/B978-0-12-817420-3.00004-7

More specifically, a subset of people with such structural impairments showed increased vulnerability to depression. Unfortunately, these structural impairments could further contribute toward additional neuronal damages.

According to a meta-review, unipolar depression was found to be associated with volume reductions in the brain areas such as the frontal cortex, cingulate cortex, orbitofrontal cortex, striatum, and hippocampus.[1] In addition, pituitary enlargement and an excess of white matter hyper intensities were reported. Brain areas such as the anterior cingulate, orbitofrontal cortex, prefrontal cortex, hippocampus, putamen, and the caudate nucleus were found associated with the pathophysiology of depression. Moreover, depression has caused volume loss in different brain areas.[2] During MDD, volume loss had been observed in different brain areas. Each one is explained in Sections 4.2.1–4.2.5.

4.2.1 Frontal Cortex

Numerous studies have reported abnormalities in the frontal cortex associated with MDD; however a detailed description would be out of scope here. A few of these studies are selected to provide the notion of volume abnormalities in the frontal cortex. For example, a reduction in prefrontal cortex volume was reported in recent literature.[3,4] In older studies, ischemic lesions were localized in the anterior frontal cortex and found to be correlated with more serious depression.[5,6] In addition, a clearly defined sequence of patients who had previously suffered from ischemic stroke showed a strong association between lesions and succeeding depression, effecting the prefronto-subcortical circuits, specifically on the left.[7] Cognitive disability was commonly observed in MDD patients, particularly involving the frontal cortex. Its volume reductions were reported as ranging from 7% overall reduction[8] to 48% in the subgenual prefrontal cortex.[9] Another evidence of notable differences from control subjects was found in diverse prefrontal cortical areas in a postmortem study of the prefrontal cortex under major depression.[10] In addition, many studies[11–14] have consistently found a volumetric loss of the complete frontal lobe and/or the OFC involving the more serious patient's groups. On the contrary, this volumetric loss could not be found in less severely ill patients.[15,16]

4.2.2 Hippocampus

Several studies have examined hippocampal volume in depression. Some[15,17–19] but not all[20,21] found significant reductions in hippocampal volumes in patients with depression. Vakili and colleagues[21] also observed correlations between depression severity and hippocampal volumes, although no group differences were reported between the depressed and control subjects. In a study, white matter abnormalities were observed; however overall differences in hippocampal volumes were not observed.[22] Moreover, studies that assessed depression severity in unipolar study participants, including those that utilized high-resolution MRI techniques, hippocampal volume loss (ranging from 8% to 19%) is commonly reported during depression.

4.2.3 Amygdala

The amygdala appears to change size based on the duration of illness. During early onset, its size may increase and afterwards starts decreasing over time. Although unipolar patients, early in the course of their illness, tend to have increased amygdala volume,[23–26] depressed patients having a longer illness duration and with a greater number of MDD episodes tend to show volumetric reductions.[13,15,16,27]

Changes involving amygdala appear to be a function of gender as female patients have shown higher amygdala reductions than their male counterparts. For example, female unipolar patients have exhibited smaller amygdalae than male patients.[16] However, some studies have reported inconsistent results, such as an increased volume reported to be associated with the right amygdala.[15] Similar findings have been reported in bilateral amygdala in first episode subjects.[22] In addition, a loss of normal asymmetry (Mervaala et al. 2000) or a reduction in the bilateral core nuclei has been observed.[28] Possible reasons for such types of abnormalities could be that the amygdala can be considered as a difficult structure to measure because the cortical amygdala merges with the surrounding cortex. In addition, the specific boundaries selected for defining amygdala are significantly varied in different studies.

4.2.4 Basal Ganglia

Multiple scientific literatures have observed a reduced volume of basal ganglia structures during major depression, specifically during late-onset depression.[29–32] Depressed patients, who were otherwise physically healthy, have shown negative results involving the caudate and putamen, a criterion less clearly found in other studies.[33] MDD patients may have abnormal basal ganglia; however it was difficult to determine the nature of the underlying changes because the characteristics of clinical samples vary from study to study.

Studies involving region of interest methods have generally been unsuccessful in determining abnormalities in basal ganglia volumes.[14,15,34]

However, previous evidence from other structural MRI studies has replicated volumetric alterations in these structures successfully.[35,36]

Shah and colleagues (2002)[14] utilized a voxel-based analysis (VBA) method and reported that treatment-resistant depressed patients had reduced caudate and putamen tissues compared to healthy controls and recovered patients. This finding implicated that the more persistent subtypes of MDD might have particular effects. In addition, this clinical sample underwent continued electroconvulsive therapy, the effects of which were largely unknown. Moreover, the hypothesis that illness progression may affect the left globus pallidus and putamen was supported by a study.[34] Hence, the opposite findings observed[15] might be illustrated by the fact that they observed a less severe patient sample than that used in other studies.[14,34] These studies suggest that the caudate nucleus, putamen, and globus pallidus may be impaired during more severe forms of depression. In short, it is difficult to draw firm conclusions about basal ganglia structures and their involvement in the neurobiology of MDD patients because of the lack of research on the basal ganglia.

4.2.5 Temporal Lobe

Studies involving temporal lobe volumes generally investigated distinct regions rather than observing the temporal lobe as a whole. Unfortunately, these studies could not find any profound abnormalities in unipolar depressed patients when compared with healthy controls.[14,15,27,37,38] However, three of these studies investigated the left and right temporal volumes separately.[11,27,38] Only Vythilingam[38] reported a lateralization effect, that is, a smaller left temporal lobe volume in patients when compared with healthy controls. Particularly, the patients observed by these researchers had a longer illness duration compared with other studies.[38] As an implication, the left-lateralized temporal lobe changes may reflect the progression of the disease over time or a distinct pathophysiological process that affects the risk of relapse.

In comparison to studies investigating the total volume of the temporal lobe, studies involving specific structures such as the basolateral temporal area and superior temporal gyrus (STG) have yielded contradictory results. For example, the study[14] reported volumetric changes in the STG of depressed patients. On the contrary, Morys and colleagues[39] failed to find such evidence. Moreover, other research efforts have identified STG volume to be inversely associated with the total length of illness and the number of depressive episodes.[14,27] This finding implies that volumetric differences may only become visible in chronically ill patients with recurrent episodes. Table 4.1 provides a summary of brain abnormalities that cause volume loss during depression.

In short, MDD patients have consistently shown abnormal frontal brain areas such as the prefrontal and anterior cingulate cortex.In addition, abnormalities in the deep cortical areas, including the hippocampus, striatum, amygdala, nucleus accumbens, and insular cortex, have also been reported during depression. In summary, Fig. 4.1 shows abnormal areas (highlighted areas) commonly reported during MDD.[40]

TABLE 4.1 Brain Abnormalities Causing Volume Loss during Depression

<table>
<tr><th>Brain Abnormalities</th><th>Affected Brain Regions</th><th>Critical Analysis</th></tr>
<tr><td>Ischemic stroke[5,6] (Modality: fMRI)</td><td>Prefrontal and subcortical circuits</td><td rowspan="5">There is no direct correlation between structural impairments in LCSPT structures and depression. Rather, it appeared that a subset of people with such structural impairment had increased vulnerability to depression and that when it did occur it may further contribute to additional damages.
There is no direct method to measure amygdala volume loss. Hence, different boundaries were defined by different researchers. The gender-specific differences were reported as a cause for contradictory findings in amygdala-related studies.</td></tr>
<tr><td>Cognitive disability[8,9] (Modality: fMRI)</td><td>Frontal cortex</td></tr>
<tr><td>Memory loss[13,15,16,27] (Modality: fMRI)</td><td>Amygdala</td></tr>
<tr><td>Memory loss[29–32] (Modality: fMRI)</td><td>Basal ganglia</td></tr>
<tr><td>Reduced BOLD responses[14,27] (Modality: fMRI)</td><td>Temporal lobe</td></tr>
</table>

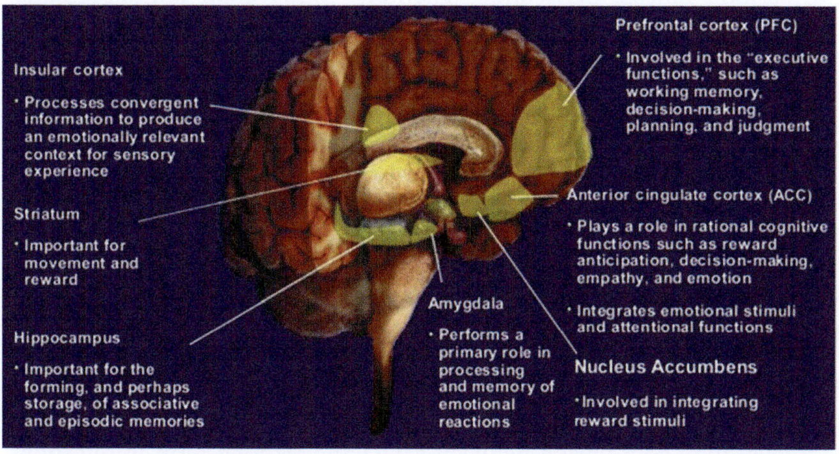

FIGURE 4.1 Abnormal brain regions during major depressive disorder.[40]

4.3 MECHANISMS UNDERLYING THE PATHOPHYSIOLOGY OF DEPRESSION

It is commonly understood that MDD is of heterogeneous nature. Therefore, the variance of the underlying pathophysiology of depression has been explained with different neurobiological theories. In Sections 4.3.1–4.3.6, a brief description of each of these factors is provided. These mechanisms include (1) genetic and nongenetic factors; (2) effects of stress hormones on depression; (3) effects of different monoamines; (4) dysfunction in specific brain regions; (5) neurotrophic hypothesis; (6) reduced GABAergic activity; (7) dysregulations of glutamate system; and/or (8) diminished circadian rhythms.

4.3.1 Genetic and Nongenetic Factors

Vulnerability to MDD can be explained through genetic and nongenetic risk factors. It is observed that genetic factors contribute between 30% and 40%, whereas nongenetic factors explain 60–70% of the vulnerability of developing MDD. These factors may include adverse childhood events of which the effects are still persistent, including childhood sexual abuse or other life trauma, low social support, marital problems, and divorce.[41] Both male and female are equally sensitive to adverse life events. However, their response may change subject to the kind of stressor. For example, men are more vulnerable to depression after having a divorce, separation, and work-related difficulties. On the other hand, women are more sensitive than men to incidents in their proximal social network such as trouble getting along with an individual, serious illness, or death.[42] Therapies other than antidepressants such as psychotherapies can be a solution for cases where MDD is caused by nongenetic factors.

4.3.2 Hypothalamic-Pituitary-Adrenal Axis Dysfunction

The hypothalamus releases the corticotropin-releasing hormone (CRH) in reaction to the perception of psychological stress by the cortical brain region. MDD is considered as a stress disorder. Most subjects treated for MDD have no evidence of dysfunction of the hypothalamic-pituitary-adrenal (HPA) axis,[43] except a few subjects who have shown abnormalities in the extra hypothalamic CRH system.[44] On the other hand, depressed participants with a background of childhood trauma have shown aberrant stress hormone secretion.[41] In addition, an increased cortisol level may be a liaison between MDD and its long-term implications such as type II diabetes, coronary heart disease, and osteoporosis.[45]

In short, HPA axis disruption may mediate abnormalities in cortisol levels and may become a source of MDD in some cases.

According to the monoamine deficiency theory, the pathophysiology of MDD includes a deficiency of neurotransmitters such as norepinephrine, serotonin, or dopamine in the central nervous system (CNS). For example, defects in the neurotransmitter/hormone system may cause the development of MDD. Moreover, during MDD, the assessment of cerebrospinal fluid chemistry, neuroendocrine changes because of pharmacotherapy, and neuroreceptor and transporter binding have demonstrated several abnormalities in the serotonergic, noradrenergic, and other neuropeptide systems. In addition, serotonin and norepinephrine are controlled by biochemical effects associated with different antidepressants.

Monoaminergic neurotransmitter systems are commonly studied for MDD. The monoaminergic neurotransmitter system is spread over the brain and is commonly found in the limbic and anterior subgenual cortex region. Serotonin, norepinephrine (vigilant concentration), and dopamine (cognitive alertness) are monoamine neurotransmitters. The endocrine system is a collection of glands that secrete hormones based on their interaction with the neurotransmitter systems. Hypothalamus is the portion in the brain that contains cells responsible for various functions; its most important task is to make connections between the nervous system and the endocrine system.

4.3.3 Neurotrophic Factors of Depression

The risk factors or vulnerabilities to having a depressive episode keep changing during MDD. For example, the first episode is generally "reactive," that is, stimulated by profound physiological stressors such as death of a loved one or a severe accident. The subsequent depressive episodes become vulnerable to minor stressors.[46] For example, any sad news on television such as a road accident may mediate a second depressive episode. During MDD, the hippocampus has been less consistently reported with a reduced volume,[47] suggesting increased stress sensitivity[48] and increased risk of reoccurrence.[49]

4.3.4 Defects in Intracellular Signaling Pathways

Cellular signaling pathways communicate at different levels, thereby establishing complex networks that permit neurons to accept and process information. These signaling pathways are certainly part of the neuroplastic phenomenon that manages complex cognitive and psychological operations as well as numerous vegetative operations such as wakefulness and appetite. Lithium and antidepressants influence cellular pathways that can manage cell survival and neuroplasticity.

Fig. 4.2 shows various influences on cellular resilience and neuroplasticity during mood disorders.[50] The development of depression might effect cellular resilience, and may cause volumetric contractions, and cell death/atrophy. Normally, these symptoms are noticed during mood disorders. Stress and depression probably cause abnormalities of cellular resilience through various mechanisms, including contractions in the brain-derived neurotrophic factor (BDNF) levels, by supporting glutamatergic transmission through *N*-methyl-D-aspartate (NMDA) and non-NMDA receptors, and by minimizing cells' energy levels.

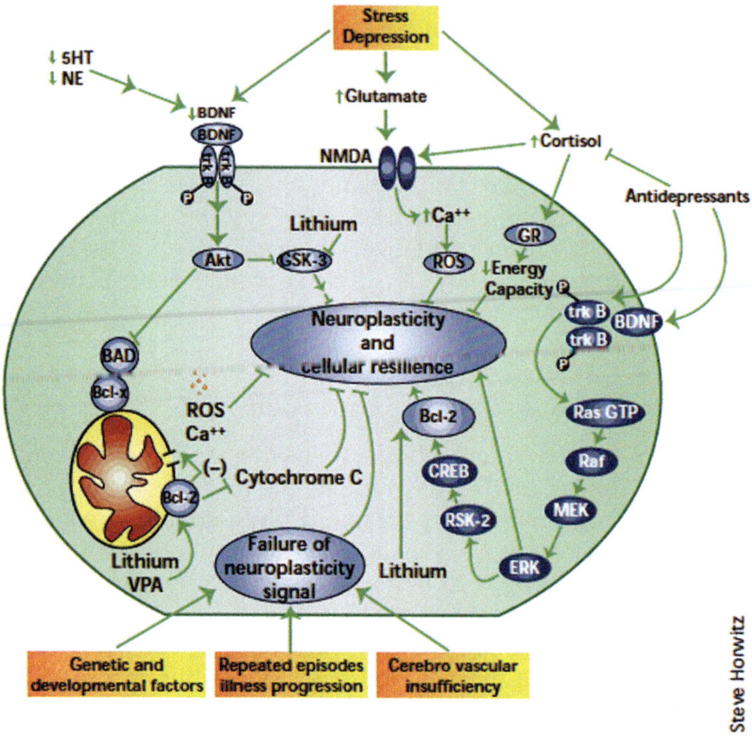

FIGURE 4.2 The cellular neurobiology of depression[50]: cyclic AMP responsive element binding protein (CREB); glucocorticoid receptor (GR); glucogen synthase kinase (GSK-3). Reactive oxygen species (ROS); Bcl-2 and Bcl-x—antiapoptotic members of the Bcl-2 family; BAD and BAX—proapoptotic members of the Bcl-2 family; nerve growth factor (NGF); ribosomal S-6 kinase (Rsk-2.); tyrosine kinase receptor for BDNF (trkB); Ras, Raf, MEK, ERK, components of the ERK–MAP-kinase pathway; brain-derived neurotrophic receptor (BDNF); pregenual anterior cingulate (PAC), subgenual prefrontal cortex (SGPFC), dorsomedial/dorsolateral and dorso anterolateral anterior cortex (DM/DALPFC), ventrolateral prefrontal cortex (VLPFC), ventral striatum.

Neurotrophic factors such as BDNF increase cell survival by mediating two different signaling pathways: the ERK–MAP-kinase pathway and the PI-3–kinase pathway. One of the main mechanisms through which BDNF boost cell survival is by enhancing the expression of the major cytoprotective protein Bcl-2. Bcl-2 diminishes cell death through various mechanisms, including damaging the excretion of calcium and cytochrome, confiscating preforms of death-inducing caspase enzymes, and increasing mitochondrial calcium uptake. The persistent administration of various types of antidepressants enhances the expression of BDNF and its receptor TrkB. Lithium and VPA strongly upregulate the cytoprotective protein Bcl-2. In addition, lithium and VPA impede GSK-3β, which induces biochemical effects having neuroprotective nature. VPA also stimulates the ERK–MAP-kinase pathway inducing effects that may have neurotrophic effects and stimulates the neurite outgrowth. Operations that cause volume reductions involve glucocorticoid neurotoxicity, lessened brain-derived neurotrophic factor (BDNF), diminished neurogenesis, and loss of plasticity.

4.3.5 Altered Glutamatergic and GABAergic Neurotransmission

According to the GABA hypothesis of depression, concentrations of GABA were found in the frontal and occipital cortex during depression.[51] The concentration of GABA found was due to critical stress effects because psychological stress appears to influence presynaptic downregulation of prefrontal GABAergic neurotransmission.[52] There was evidence that acute stress may decrease GABA-A receptor operation by altering neuroactive steroid synthesis.[53]

4.3.6 Circadian Rhythms

Table 4.2 shows the diagnostic criteria of MDD that include sleep disturbances and daytime fatigue. This indicates abnormal sleep–wake regulation during depression. A group of depressed people have exhibited circadian rhythm disorder.[54]

4.4 ELECTROENCEPHALOGRAPHY CORRELATES FOR DEPRESSION

As electroencephalography (EEG) data are considered direct representations of underlying brain activities, researchers have reported

TABLE 4.2 Mechanisms Underlying Pathophysiology of Depression

Abnormal Neural Mechanism	Proposed Causes/Effects	Critical Analysis
Genetic and nongenetic factors[41,42]	Stressors such as life trauma, low social support, marital problems, and more.	HPA axis dysfunction may mediate abnormalities in cortisol levels and may become a source of MDD in some cases. However, HPA axis dysfunction is mainly associated with patients affected by acute stress.
HPA axis dysfunction[43–45]	Corticotropin-releasing hormones.	
Hormonal deficiencies	Deficiency of neurotransmitter norepinephrine, serotonin, or dopamine in the central nervous system (CNS).	
Neurotrophic factors of depression[46–49]	News of the death of a loved one. Later, any sad news such as a road accident may mediate a second depressive episode.	
Intracellular signaling pathways[50]	Neuroplasticity and cellular resilience contribute toward the development of mood disorders.	
Altered glutamatergic-GABAergic neurotransmission[51,52]	Concentration of GABA in frontal and occipital cortex was found during depression.	
Disturbances in circadian rhythms[54,55]	Sleep disturbances and daytime fatigue could be either causes or effects during depression.	

MDD-specific brain regions based on EEG analysis only. These findings were observed while comparing a group of MDD patients with healthy controls. For example, changes in EEG frequency bands that are specific to certain brain regions and exhibited by the MDD group only were reported. In this context, a study[56] has reported an increase in the theta and delta power over the right hemisphere of the brain observed in 20 depressed patients. However, this observation could not be found in healthy controls.

Low resolution electromagnetic tomography (LORETA)-based studies have revealed that the anterior cingulate cortex (ACC),[57] lateral orbitofrontal cortex, and dorsolateral prefrontal cortex (DLPFC)[58] are correlated with depression. A further description of EEG-based abnormalities has been shown in Table 4.3. The table summarizes EEG changes or

TABLE 4.3 Major Depressive Disorder -Specific EEG Changes and Related Brain Regions

MDD-Specific Brain Regions	Changes in EEG Bands Specific to MDD
Right hemisphere abnormalities[56]	Elevated theta over the right hemisphere was observed only in depressed patients when compared with healthy controls.
Anterior cingulate cortex (ACC)[57]	LORETA analysis disclosed increased delta, theta, and beta power.
ACC, DLPFC, lateral orbitofrontal cortex[58]	LORETA analysis revealed increased alpha power in MDD patients.
Parietal and occipital electrode sites[59]	MDD patients show elevated theta and alpha activity.
Left frontal region,[60] DLPFC[61]	Left frontal alpha activity was reported in depressed patients when compared with controls.
Frontal region cortical activity[62,63]	Depression has been characterized as functional asymmetric cortical activity.
Frontal regions[64]	Depressed patients have shown frontal alpha asymmetry.
Posterior and temporoparietal regions[65,66]	Depressed patients have shown alpha asymmetry in these regions with relatively lessened right cortical activation.
Right anterior and left posterior regions[67]	Depressed patients have shown impaired functional connectivity at EEG alpha and theta bands.

patterns that have been associated with MDD-specific brain regions commonly manifested in depressed patients.

It is important to note that EEG and event-related potentials (ERP) changes related to MDD could be utilized for diagnosis and treatment of MDD.

For example, abnormal alpha and theta band activities have shown association with cognitive and memory performance, which could be associated with the hippocampus. As mentioned in Tables 4.2 and 4.3, the hippocampus has been seen to be affected during MDD which implies that a depressed individual may show abnormal cognitive and memory performance. This finding may help in the effective treatment of depressed patients, for example, the theta and alpha ratio could be improved by adopting neurofeedback techniques that may ultimately improve MDD.[68]

The administration of antidepressants such as SSRIs has profound positive effects on the human brain and could be observed with EEG data acquired from MDD patients under pharmacological treatment. For example, an elevation in theta and reduction in alpha activity has been observed in healthy individuals while buspirone was administered.[69–72] In addition, the localization of buspirone-related effects with LORETA has revealed a profound elevation in theta activity in the hippocampus and neighboring cortical areas. These findings have suggested that EEG changes of power in theta bands could be utilized to study the effects of antidepressants on the human brain.

4.5 ELECTROENCEPHALOGRAPHY-BASED DIAGNOSIS OF DEPRESSION

Abnormal neuronal activities were found in MDD patients' EEG and ERP data when compared with healthy controls. For example, differences in power spectrum, lateralization of power asymmetries, and larger intensities or lower amplitudes of P300 components were found in MDD patients. On the other hand, some studies utilized machine learning (ML) techniques to automatically detect aberrant EEG/ERP patterns associated with the disease' conditions. Sections 4.5.1–4.5.3 describe studies based on EEG- and ERP-derived features to differentiate between depressed patients and healthy controls. Such investigations have provided a basis for diagnosing MDD using EEG and ERP.

4.5.1 Electroencephalography Frequency Bands

The EEG signal is of composite nature and can be decomposed into its constituent frequency bands such as delta (0.5–4 Hz), theta (5–8 Hz), alpha (8–12 Hz), beta (12–30 Hz), and gamma (frequencies greater than 30 Hz). Power computations of particular frequency bands have shown association with the pathophysiology of depression. For example, in an earlier study, elevated EEG activity (less alpha) was observed during resting condition[73–76] and increased relative power was also reported.[77,78] The elevation in EEG activity was found to involve the parietal,[79] occipital,[80] and frontal brain regions. Moreover, the early stages of depression were identified by increased alpha activity.[59] However, some studies could not observe alpha activations as significantly different between MDD patients and healthy controls.[81,82] Increased beta band activations were reported in depressed individuals.[83,84]

EEG alpha interhemispheric asymmetry has been studied as a vulnerability marker for depression. Davidson and colleagues[85], had hypothesized that the depression could cause hypoactive left and hyperactive right prefrontal cortex. In 1983, the researchers found a relative hyperactivation of the right prefrontal cortex.[85] In a later study, the researchers' supposed "approach" and "withdrawal" were as orthogonal to each other and as fundamental to EEG asymmetry, which may become a vulnerability measure for depression.[86] The approach system facilitated appetitive behavior with positive results and the withdrawal system motivated aversive and negative emotions.[87] Reduced left-sided frontal EEG activation was associated with a decrease in the approach system. These findings implicated that study participants with such symptoms were at risk of negative emotional states and depression in response to environmental stress.

Unfortunately, studies based on EEG alpha interhemispheric asymmetry have exhibited conflicting implications. For example, in a recent study, asymmetry was observed to be profoundly increased in depressed patients than the healthy controls.[88] However, other studies[89,90] reported a decreased right frontal activity relative to the left side.

4.5.2 Event-Related Potentials Component: P300

The ERP component (P300) had been associated with cognitive abilities and is commonly studied for MDD as well. For example, auditory-evoked potentials (AEP) have shown positive associations with cognitive abilities.[91,92]

Regarding EEG frequency bands, depression has been associated with elevated EEG activity (less alpha). Elevated alpha was associated with antidepressant treatment response. In the context of EEG alpha asymmetry, hypoactive left frontal cortex and hyperactive right frontal cortex were observed during depression. Decreased left-sided frontal activations can be a vulnerability indicator for depression. With focus on the ERP component such as P300, depressed participants manifested larger P300 latency and smaller P300 amplitudes compared to healthy controls. Further critical analysis is provided in Table 4.4.

In general, MDD patients were considered to have low cognitive abilities due to the illness. Such abnormalities were observed with a change in P300 intensity and the occurrence or latency of a P300 peak. For example, depression was associated with a delay in the occurrence of a P300 peak[93] and only found in MDD patients when compared with healthy controls.[94–96] In addition, a decreased P300 intensity in the right hemisphere was observed based on the LORETA analysis.[97] Moreover, longer P300 latency was observed in a study involving visually evoked stimuli.[98]

TABLE 4.4 EEG/ERP Data to Discriminate Major Depressive Disorder Patients from Healthy Controls

Brain Dynamics	Fundamental Results	Comments and Critical Analysis
EEG frequency bands[73–76] (activations in alpha and beta bands)	Depression has been associated with elevated EEG activity (less alpha). Elevated alpha was associated with the response to antidepressant treatment.	The aberrant patterns observed in different EEG bands have shown promises during depression. However, they were not specific to either the patients or controls. Hence, interstudy comparisons are mandatory to enhance future research outcomes and to formalize methodological procedures.
EEG alpha asymmetry[85–87,89,90] (changes in alpha activation between left and right hemispheres)	Hyperactive right frontal cortex and hypoactive left frontal cortex manifested during depression. Decreased left-sided frontal activations can be a vulnerability indicator for depression.	Factors such as increasing the sample sizes and balancing the gender distributions of the studies involving EEG alpha asymmetry could improve low specificity of the findings. In order to have common evaluation criteria, the results should include the classification sensitivity and specificity values.
P300[91–98] (P300 amplitudes and delayed P300 peak)	In general, depressed patients have exhibited larger P300 latency and smaller P300 amplitudes when compared with healthy controls.	Many researchers believed that P300 is an index of cognition. However, not all patients have exhibited common changes in the latency and amplitudes of P300 because of the administration of antidepressants.

4.5.3 Machine Learning Methods to Diagnose Depression

Automatic selection of EEG patterns specific to MDD is possible using ML techniques. EEG data mining has been gaining importance in diagnosing various mental illnesses.[99] EEG data in combination with ML techniques have been utilized for diagnosing MDD.[100–102] These studies utilized EEG-based features such as Katz's and Higuchi's fractal dimension (HFD).[100] In addition, a combination of spectral asymmetry and HFD were utilized as well.[103] In addition to EEG, ERP features such as P600 have shown association with MDD.[104]

In summary, these features derived from EEG and ERP data have shown promising results in diagnosing depression. However, due to limitations associated with the studies, their findings cannot be translated into clinical applications. The contradictions in the findings of various studies have posed major questions for the development of objective methods that could be confidently used in psychiatry clinics. Table 4.5 provides a detailed description of the ML method for the diagnosis of depression.

TABLE 4.5 ML Methods for Diagnosing Depression

Study	Year	Sample Size	Classification Results
EEG power, frequency, asymmetry, and coherence in depressed males[83]	2001	70 MDD 23 Controls	Accuracy = 91.3%
Evaluation of an SVM-based computer-aided diagnosis (CAD) system to classify MDD patients from healthy controls involving the P600 ERP signal component[104]	2004	25 MDD 25 Controls	Accuracy = 94%
EEG features such as spectral asymmetry and Higuchi's fractal dimension[103]	2010	25 MDD 25 Controls	True detection rate of 88% for patients and 82% for controls
Diagnosis of psychiatric disorders using EEG data and applying a statistical decision model[105]	2010	64 MDD 207 Controls	Accuracy = 85%
Unsupercised classification of MDD using functional connectivity MRI[106]	2014	24 MDD 29 Controls	Accuracy = 92.5%
Computer-aided diagnosis of depression using EEG signals[107]	2015	25 MDD 25 Controls	Accuracy = 91%
Classifying depressed patients and normal subjects using machine learning techniques and nonlinear features from EEG signals[101]	2015	45 MDD 45 Controls	Accuracy = 90%
Data mining EEG signals in depression for their diagnostic value[108]	2015	53 MDD 43 Controls	Accuracy = 80%

4.6 SUMMARY

This chapter elaborates on the pathophysiology of depression from different perspectives such as pathophysiology caused by brain volume abnormalities or various other mechanisms that could become a cause or effect of depression. In addition, the pathophysiology of depression has been explained from an EEG perspective. For example, different correlates of EEG with depression have been discussed. EEG-based findings form the basis for automatic diagnosis of depression termed as EEG-based diagnosis. This last point shall be further elaborated in detail in the subsequent chapters.

References

[1] Arnone D, McIntosh A, Ebmeier K, Munafo M, Anderson I. Magnetic resonance imaging studies in unipolar depression: systematic review and meta-regression analyses. *Eur Neuropsychopharmacol*. 2012;22:1–16.

[2] Koolschijn P, van Haren NE, Lensvelt-Mulders GJ, Pol H, Hilleke E, Kahn RS. Brain volume abnormalities in major depressive disorder: a meta-analysis of magnetic resonance imaging studies. *Hum Brain Mapp*. 2009;30:3719–3735.

[3] Salvadore G, Nugent AC, Lemaitre H, et al. Prefrontal cortical abnormalities in currently depressed versus currently remitted patients with major depressive disorder. *Neuroimage*. 2011;54:2643–2651.

[4] Treadway MT, Waskom ML, Dillon DG, et al. Illness progression, recent stress, and morphometry of hippocampal subfields and medial prefrontal cortex in major depression. *Biol Psychiatry*. 2014;77(3):285–294.

[5] Robinson RG, Kubos KL, Starr LB, Rao K, Price TR. Mood changes in stroke patients: relationship to lesion location. *Compr Psychiatry*. 1983;24:555–566.

[6] Lipsey JR, Robinson R, Pearlson G, Rao K, Price T. Mood change following bilateral hemisphere brain injury. *Br J Psychiatry*. 1983;143:266–273.

[7] Vataja R, Pohjasvaara T, Leppävuori A, et al. Magnetic resonance imaging correlates of depression after ischemic stroke. *Arch Gen Psychiatry*. 2001;58:925–931.

[8] Coffey C, Wilkinson W, Parashos L, et al. Quantitative cerebral anatomy of the aging human brain: a cross-sectional study using magnetic resonance imaging. *Neurology*. 1992;42. pp. 527–527.

[9] Drevets WC, Price JL, Simpson JR, et al. Subgenual prefrontal cortex abnormalities in mood disorders. *Nature*. 1997;386:824–827.

[10] Rajkowska G, Miguel-Hidalgo JJ, Wei J, et al. Morphometric evidence for neuronal and glial prefrontal cell pathology in major depression. *Biol Psychiatry*. 1999;45:1085–1098.

[11] Bremner JD, Vythilingam M, Vermetten E, et al. Reduced volume of orbitofrontal cortex in major depression. *Biol Psychiatry*. 2002;51:273–279.

[12] Lacerda AL, Keshavan MS, Hardan AY, et al. Anatomic evaluation of the orbitofrontal cortex in major depressive disorder. *Biol Psychiatry*. 2004;55:353–358.

[13] Monkul E, Hatch JP, Nicoletti MA, et al. Fronto-limbic brain structures in suicidal and non-suicidal female patients with major depressive disorder. *Mol Psychiatry*. 2006;12:360–366.

[14] Shah P, Glabus M, Goodwin G, Ebmeier K. Chronic, treatment-resistant depression and right fronto-striatal atrophy. *Br J Psychiatry*. 2002;180:434–440.

[15] Bremner JD, Narayan M, Anderson ER, Staib LH, Miller HL, Charney DS. Hippocampal volume reduction in major depression. *Am J Psychiatry.* 2000;157:115–118.

[16] Hastings RS, Parsey RV, Oquendo MA, Arango V, Mann JJ. Volumetric analysis of the prefrontal cortex, amygdala, and hippocampus in major depression. *Neuropsychopharmacology.* 2004;29:952–959.

[17] Elbejjani M, Fuhrer R, Abrahamowicz M, et al. Hippocampal atrophy and subsequent depressive symptoms in older men and women: results from a 10-year prospective cohort. *Am J Epidemiol.* 2014;180:385–393.

[18] Opel N, Redlich R, Zwanzger P, et al. Hippocampal atrophy in major depression: a function of childhood maltreatment rather than diagnosis and quest. *Neuropsychopharmacology.* 2014;39:2723–2731.

[19] Stratmann M, Konrad C, Kugel H, et al. Insular and hippocampal gray matter volume reductions in patients with major depressive disorder. *PLoS ONE.* 2014;9: e102692.

[20] Mervaala E, Föhr J, Könönen M, et al. Quantitative MRI of the hippocampus and amygdala in severe depression. *Psychol Med.* 2000;30:117–125.

[21] Vakili K, Pillay SS, Lafer B, et al. Hippocampal volume in primary unipolar major depression: a magnetic resonance imaging study. *Biol Psychiatry.* 2000;47:1087–1090.

[22] Frodl T, Meisenzahl EM, Zetzsche T, et al. Hippocampal changes in patients with a first episode of major depression. *Am J Psychiatry.* 2002;159:1112–1118.

[23] Frodl T, Meisenzahl E, Zetzsche T, et al. Enlargement of the amygdala in patients with a first episode of major depression. *Biol Psychiatry.* 2002;51:708–714.

[24] Frodl T, Meisenzahl EM, Zetzsche T, et al. Larger amygdala volumes in first depressive episode as compared to recurrent major depression and healthy control subjects. *Biol Psychiatry.* 2003;53:338–344.

[25] Lange C, Irle E. Enlarged amygdala volume and reduced hippocampal volume in young women with major depression. *Psychol Med.* 2004;34:1059–1064.

[26] Weniger G, Lange C, Irle E. Abnormal size of the amygdala predicts impaired emotional memory in major depressive disorder. *J Affect Disord.* 2006;94:219–229.

[27] Caetano SC, Hatch JP, Brambilla P, et al. Anatomical MRI study of hippocampus and amygdala in patients with current and remitted major depression. *Psychiatry Res. Neuroimaging.* 2004;132:141–147.

[28] Sheline YI, Gado MH, Price JL. Amygdala core nuclei volumes are decreased in recurrent major depression. *Neuroreport.* 1998;9:2023–2028.

[29] Greenwald B, Kramer-Ginsberg E, Bogerts B, et al. Qualitative magnetic resonance imaging findings in geriatric depression. Possible link between later-onset depression and Alzheimer's disease? *Psychol Med.* 1997;27:421–431.

[30] Husain MM, McDonald WM, Doraiswamy PM, et al. A magnetic resonance imaging study of putamen nuclei in major depression. *Psychiatry Res. Neuroimaging.* 1991;40:95–99.

[31] Krishnan KRR, McDonald WM, Escalona PR, et al. Magnetic resonance imaging of the caudate nuclei in depression: preliminary observations. *Arch Gen Psychiatry.* 1992;49:553–557.

[32] Steffens DC, Krishnan K. Structural neuroimaging and mood disorders: recent findings, implications for classification, and future directions. *Biol Psychiatry.* 1998;43:705–712.

[33] Lenze EJ, Sheline YI. Absence of striatal volume differences between depressed subjects with no comorbid medical illness and matched comparison subjects. *Am J Psychiatry.* 1999;156:1989–1991.

[34] Lacerda AL, Nicoletti MA, Brambilla P, et al. Anatomical MRI study of basal ganglia in major depressive disorder. *Psychiatry Res Neuroimaging.* 2003;124:129–140.

[35] Bonelli R, Kapfhammer H-P, Pillay S, Yurgelun-Todd D. Basal ganglia volumetric studies in affective disorder: what did we learn in the last 15 years? *J Neural Transm.* 2006;113:255–268.

[36] Hickie IB, Naismith SL, Ward PB, et al. Serotonin transporter gene status predicts caudate nucleus but not amygdala or hippocampal volumes in older persons with major depression. *J Affect Disord.* 2007;98:137–142.

[37] Frodl T, Meisenzahl EM, Zill P, et al. Reduced hippocampal volumes associated with the long variant of the serotonin transporter polymorphism in major depression. *Arch Gen Psychiatry.* 2004;61:177–183.

[38] Vythilingam M, Vermetten E, Anderson GM, et al. Hippocampal volume, memory, and cortisol status in major depressive disorder: effects of treatment. *Biol Psychiatry.* 2004;56:101–112.

[39] Moryś JM, Bobek-Billewicz B, Dziewiątkowski J, et al. A magnetic resonance volumetric study of the temporal lobe structures in depression. *Folia Morphol (Praha).* 2003;62:347–352.

[40] Charney DS, Buxbaum JD, Sklar P, Nestler EJ. *Neurobiology of Mental Illness.* Oxford University Press; 2013.

[41] Heim C, Newport DJ, Mletzko T, Miller AH, Nemeroff CB. The link between childhood trauma and depression: insights from HPA axis studies in humans. *Psychoneuroendocrinology.* 2008;33:693–710.

[42] Kendler KS, Thornton LM, Prescott CA. Gender differences in the rates of exposure to stressful life events and sensitivity to their depressogenic effects. *Am J Psychiatry.* 2001;158:587–593.

[43] Belmaker R, Agam G. Major depressive disorder. *N Engl J Med.* 2008;358:55–68.

[44] Pariante CM, Lightman SL. The HPA axis in major depression: classical theories and new developments. *Trends Neurosci.* 2008;31:464–468.

[45] Gold PW, Chrousos GP. The endocrinology of melancholic and atypical depression: relation to neurocircuitry and somatic consequences. *Proc Assoc Am Physicians.* 1999;111:22–34.

[46] Kessing LV, Hansen MG, Andersen PK, Angst J. The predictive effect of episodes on the risk of recurrence in depressive and bipolar disorders—a life-long perspective. *Acta Psychiatr Scand.* 2004;109:339–344.

[47] Sheline YI, Gado MH, Kraemer HC. Untreated depression and hippocampal volume loss. *Am J Psychiatry.* 2003;160:1516–1518.

[48] Hasler G, Fromm S, Alvarez RP, Luckenbaugh DA, Drevets WC, Grillon C. Cerebral blood flow in immediate and sustained anxiety. *J Neurosci.* 2007;27:6313–6319.

[49] Frodl TS, Koutsouleris N, Bottlender R, et al. Depression-related variation in brain morphology over 3 years: effects of stress? *Arch Gen Psychiatry.* 2008;65:1156–1165.

[50] Manji HK, Drevets WC, Charney DS. The cellular neurobiology of depression. *Nat Med.* 2001;7:541–547.

[51] Hasler G, van der Veen JW, Tumonis T, Meyers N, Shen J, Drevets WC. Reduced prefrontal glutamate/glutamine and γ-aminobutyric acid levels in major depression determined using proton magnetic resonance spectroscopy. *Arch Gen Psychiatry.* 2007;64:193–200.

[52] Hasler G, van der Veen JW, Grillon C, Drevets WC, Shen J. Effect of acute psychological stress on prefrontal GABA concentration determined by proton magnetic resonance spectroscopy. *Am J Psychiatry.* 2010;167:1226–1231.

[53] Rajkowska G, O'Dwyer G, Teleki Z, Stockmeier CA, Miguel-Hidalgo JJ. GABAergic neurons immunoreactive for calcium binding proteins are reduced in the prefrontal cortex in major depression. *Neuropsychopharmacology.* 2006;32:471–482.

[54] Bunney WE, Bunney BG. Molecular clock genes in man and lower animals: possible implications for circadian abnormalities in depression. *Neuropsychopharmacology.* 2000;22:335–345.

[55] Hasler G, Drevets WC, Manji HK, Charney DS. Discovering endophenotypes for major depression. *Neuropsychopharmacology*. 2004;29:1765–1781.

[56] Kwon JS, Youn T, Jung HY. Right hemisphere abnormalities in major depression: quantitative electroencephalographic findings before and after treatment. *J Affect Disord*. 1996;40:169–173.

[57] Korb AS, Hunter AM, Cook IA, Leuchter AF. Rostral anterior cingulate cortex theta current density and response to antidepressants and placebo in major depression. *Clin Neurophysiol*. 2009;120:1313–1319.

[58] Korb AS, Cook IA, Hunter AM, Leuchter AF. Brain electrical source differences between depressed subjects and healthy controls. *Brain Topogr*. 2008;21:138–146.

[59] Grin-Yatsenko VA, Baas I, Ponomarev VA, Kropotov JD. Independent component approach to the analysis of EEG recordings at early stages of depressive disorders. *Clin Neurophysiol*. 2010;121:281–289.

[60] Davidson RJ, Henriques J, "Regional brain function in sadness and depression," 2000.

[61] Herrington JD, Heller W, Mohanty A, et al. Localization of asymmetric brain function in emotion and depression. *Psychophysiology*. 2010;47:442–454.

[62] Allen JJ, Urry HL, Hitt SK, Coan JA. The stability of resting frontal electroencephalographic asymmetry in depression. *Psychophysiology*. 2004;41:269–280.

[63] Allen J. The state and trait nature of frontal EEG asymmetry in emotion. *Psychophysiology*. 2000. pp. S18-S18.

[64] Debener S, Beauducel A, Nessler D, et al. Is resting anterior EEG alpha asymmetry a trait marker for depression? *Neuropsychobiology*. 2000;41:31–37.

[65] Bruder GE, Fong R, Tenke CE, et al. Regional brain asymmetries in major depression with or without an anxiety disorder: a quantitative electroencephalographic study. *Biol Psychiatry*. 1997;41:939–948.

[66] Kentgen LM, Tenke CE, Pine DS, Fong R, Klein RG, Bruder GE. Electroencephalographic asymmetries in adolescents with major depression: influence of comorbidity with anxiety disorders. *J Abnorm Psychol*. 2000;109:797.

[67] Fingelkurts AA, Fingelkurts AA, Rytsälä H, Suominen K, Isometsä E, Kähkönen S. Impaired functional connectivity at EEG alpha and theta frequency bands in major depression. *Hum Brain Mapp*. 2007;28:247–261.

[68] Gruzelier J. A theory of alpha/theta neurofeedback, creative performance enhancement, long distance functional connectivity and psychological integration. *Cogn Process*. 2009;10:101–109.

[69] Anderer P, Saletu B, Pascual-Marqui RD. Effect of the 5-HT 1A partial agonist buspirone on regional brain electrical activity in man: a functional neuroimaging study using low-resolution electromagnetic tomography (LORETA). *Psychiatry Res. Neuroimaging*. 2000;100:81–96.

[70] Barbanoj M, Anderer P, Antonijoan R, Torrent J, Saletu B, Jané F. Topographic pharmaco-EEG mapping of increasing doses of buspirone and its comparison with diazepam. *Hum Psychopharmacol*. 1994;9:101–109.

[71] Holland R, Wesnes K, Dietrich B. Single dose human pharmacology of umespirone. *Eur J Clin Pharmacol*. 1994;46:461–468.

[72] McAllister-Williams RH. Misinterpretation of randomized trial evidence: do antidepressants work? *Brit J Hosp Med (Lond)*. 2008;69:246–247.

[73] Lemere F. The significance of individual differences in the Berger rhythm. *Brain*. 1936;59:366–375.

[74] Begić D, Popović-Knapić V, Grubišin J, et al. Quantitative electroencephalography in schizophrenia and depression. *Psychiatr Danub*. 2011;23:355–362.

[75] Jaworska N, Blier P, Fusee W, Knott V. Alpha power, alpha asymmetry and anterior cingulate cortex activity in depressed males and females. *J Psychiatr Res*. 2012;46:1483–1491.

[76] Roemer RA, Shagass C, Dubin W, Jaffe R, Siegal L. Quantitative EEG in elderly depressives. *Brain Topogr*. 1992;4:285–290.

[77] John ER, Prichep L, Fridman J, Easton P. Neurometrics: computer-assisted differential diagnosis of brain dysfunctions. *Science*. 1988;239:162–169.

[78] Prichep LS, John E. QEEG profiles of psychiatric disorders. *Brain Topogr*. 1992;4:249–257.

[79] Grin-Yatsenko VA, Baas I, Ponomarev VA, Kropotov JD. EEG power spectra at early stages of depressive disorders. *J Clin Neurophysiol*. 2009;26:401–406.

[80] Bruder GE, Sedoruk JP, Stewart JW, McGrath PJ, Quitkin FM, Tenke CE. Electroencephalographic alpha measures predict therapeutic response to a selective serotonin reuptake inhibitor antidepressant: pre- and post-treatment findings. *Biol Psychiatry*. 2008;63:1171–1177.

[81] Flor-Henry P, Koles Z, Howarth B, Burton L. Neurophysiological studies of schizophrenia, mania and depression. In: Gruzelier J, Flor-Henry P, eds. *Hemisphere Asymmetries of Function in Psychopathology*. Amsterdam: Elsevier; 1979:189–221.

[82] Knott VJ, Lapierre YD. Computerized EEG correlates of depression and antidepressant treatment. *Prog Neuropsychopharmacol Biol Psychiatry*. 1987;11:213–221.

[83] Knott V, Mahoney C, Kennedy S, Evans K. EEG power, frequency, asymmetry and coherence in male depression. *Psychiatry Res.Neuroimaging*. 2001;106:123–140.

[84] Lieber AL, Prichep LS. Diagnosis and subtyping of depressive disorders by quantitative electroencephalography: I. Discriminant analysis of selected variables in untreated depressives. *Hillside J Clin Psychiatry*. 1988;10:71–83.

[85] Schaffer CE, Davidson RJ, Saron C. Frontal and parietal electroencephalogram asymmetry in depressed and nondepressed subjects. *Biol Psychiatry*. 1983;18(7):753–762.

[86] Henriques JB, Davidson, RJ. Regional brain electrical asymmetries discriminate between previously depressed and healthy control subjects. *J Abnorm Psychol*. 1990;99:22–31.

[87] Davidson RJ. Anterior electrophysiological asymmetries, emotion, and depression: conceptual and methodological conundrums. *Psychophysiology*. 1998;35:607–614.

[88] Gollan JK, Hoxha D, Chihade D, Pflieger ME, Rosebrock L, Cacioppo J. Frontal alpha EEG asymmetry before and after behavioral activation treatment for depression. *Biol Psychol*. 2014;99:198–208.

[89] Chang JS, Yoo CS, Yi SH, et al. An integrative assessment of the psychophysiologic alterations in young women with recurrent major depressive disorder. *Psychosom Med*. 2012;74:495–500.

[90] Flor-Henry P. Lateralized temporal limbic dysfunction and psychopathology. *Ann N Y Acad Sci*. 1976;280:777–795.

[91] Mulert C, Pogarell O, Juckel G, et al. The neural basis of the P300 potential. *Eur Arch Psychiatry Clin Neurosci*. 2004;254:190–198.

[92] Volpe U, Mucci A, Bucci P, Merlotti E, Galderisi S, Maj M. The cortical generators of P3a and P3b: a LORETA study. *Brain Res Bull*. 2007;73:220–230.

[93] Bruder GE, Towey JP, Stewart JW, Friedman D, Tenke C, Quitkin FM. Event-related potentials in depression: influence of task, stimulus hemifield and clinical features on P3 latency. *Biol Psychiatry*. 1991;30:233–246.

[94] Blackwood D, Sharp C, Walker M, Doody G, Glabus M, Muir W. Implications of comorbidity for genetic studies of bipolar disorder: P300 and eye tracking as biological markers for illness. *Br J Psychiatry Suppl*. 1996;85–92.

[95] O'donnell B, Vohs J, Hetrick W, Carroll C, Shekhar A. Auditory event-related potential abnormalities in bipolar disorder and schizophrenia. *Int J Psychophysiol*. 2004;53:45–55.

[96] Souza VB, Muir WJ, Walker MT, et al. Auditory P300 event-related potentials and neuropsychological performance in schizophrenia and bipolar affective disorder. *Biol Psychiatry*. 1995;37:300–310.

[97] Kawasaki T, Tanaka S, Wang J, Hokama H, Hiramatsu K. Abnormalities of P300 cortical current density in unmedicated depressed patients revealed by LORETA analysis of event related potentials. *Psychiatry Clin Neurosci*. 2004;58:68–75.

[98] Zhu Y, Chen X-S, Qiu J-Y. Study on visual P300 evoked by facial expression stimulus in patients with depression. *J Shanghai Jiaotong Univ. (Med. Sci.)*. 2012;10:014.

[99] Klöppel S, Abdulkadir A, Jack Jr CR, Koutsouleris N, Mourão-Miranda J, Vemuri P. Diagnostic neuroimaging across diseases. *Neuroimage*. 2012;61:457–463.

[100] Ahmadlou M, Adeli H, Adeli A. Fractality analysis of frontal brain in major depressive disorder. *Int J Psychophysiol*. 2012;85:206–211.

[101] Hosseinifard B, Moradi MH, Rostami R. Classifying depression patients and normal subjects using machine learning techniques and nonlinear features from EEG signal. *Comput Methods Programs Biomed*. 2013;109:339–345.

[102] Zhang Z-X, Tian X-W, Lim JS, New algorithm for the depression diagnosis using HRV: a neuro-fuzzy approach. In: *Bioelectronics and Bioinformatics (ISBB), 2011 International Symposium on* Suzhou, China; 2011 :pp. 283-286.

[103] Bachmann M, Lass J, Suhhova A, Hinrikus H. Spectral asymmetry and Higuchi's fractal dimension measures of depression electroencephalogram. *Comput Math Methods Med*. 2013;2013.

[104] Kalatzis I, Piliouras N, Ventouras E, Papageorgiou CC, Rabavilas AD, Cavouras D. Design and implementation of an SVM-based computer classification system for discriminating depressive patients from healthy controls using the P600 component of ERP signals. *Comput Methods Programs Biomed*. 2004;75:11–22.

[105] Khodayari-Rostamabad A, Reilly JP, Hasey G, MacCrimmon D, Diagnosis of psychiatric disorders using EEG data and employing a statistical decision model, in *Engineering in Medicine and Biology Society (EMBC), 2010 Annual International Conference of the IEEE*. Buenos Aires, Argentina; 2010: 4006-4009.

[106] Zeng LL, Shen H, Liu L, Hu D. Unsupervised classification of major depression using functional connectivity MRI. *Hum Brain Mapp*. 2014;35:1630–1641.

[107] Acharya UR, Sudarshan VK, Adeli H, et al. A novel depression diagnosis index using nonlinear features in EEG signals. *Eur Neurol*. 2015;74:79–83.

[108] Mohammadi M, Al-Azab F, Raahemi B, et al. Data mining EEG signals in depression for their diagnostic value. *BMC Med Inform Decis Mak*. 2015;15:108.

CHAPTER

5

Using Electroencephalography for Diagnosing and Treating Depression

5.1 INTRODUCTION

This chapter provides a review regarding electroencephalography (EEG)-based studies that point to objective measures for diagnosing depression and assessing its treatment efficacy. Recently, EEG-based diagnosis of depression has emphasized the use of machine learning (ML) methods, also termed as computer-aided diagnosis (CAD)-based solutions.[1] On the other hand, EEG-based studies for antidepressant treatment efficacy assessment often provide measures that show correlation specific to either treatment responder or nonresponder groups. Details on such studies are also provided.

The literature has evidenced many EEG-based studies to quantify the differences between depressed and healthy controls.[2] Differences can be determined based on the extraction of useful EEG-based features such as EEG signal power in distinct frequency bands, EEG alpha asymmetry, and connectivity measures (coherence, phase delay, synchronization likelihood, etc.). These distinct features form the basis for developing and training models to perform classifications on new EEG data in clinical settings. This idea forms the basis of EEG-based diagnosis and treatment selection. For example, a ML model trained on EEG data has shown correlation with clinical assessment questionnaires,[3] implicating its capability to be used as a diagnosis tool. During such studies, questionnaires are considered as gold standard because questionnaires can provide information about physiological conditions and could be utilized for comparison purposes during EEG analysis.

More recently, there is growing interest in health information technology involving evidence-based approaches in clinical decision

DOI: https://doi.org/10.1016/B978-0-12-817420-3.00005-9

support systems. The purpose of this is to achieve "precision medicine" using "interventional informatics." However, evidence-based approaches show less sustainability; therefore interventional informatics using data-driven interventions is required to achieve evidence-based clinical decision support.[4]

5.2 ELECTROENCEPHALOGRAPHY-BASED DIAGNOSIS FOR DEPRESSION

From a clinical perspective, early diagnosis of depression has fundamental importance in treating this condition, as early diagnosis can decrease the incidence of mortality in patients with depression. Hence, improvement in the quality of life of depressed patients could be possible.

5.2.1 Electroencephalography Frequency Bands

The EEG signal is composite in nature and decomposes into distinct frequency bands such as delta, theta, alpha, beta, and gamma. Therefore, the EEG features can be computed in the frequency domain. For example, the EEG power spectrum, interhemispheric asymmetry (the mathematical formula for asymmetry is based on the computation of power), and coherence (coherence is a frequency-domain counterpart of correlation) are computed involving different frequency bands. Researchers have computed signal power, asymmetry, and coherence to investigate the differences of these measures among depressed male and female subjects.[5] The study examined the utility of EEG-based absolute and relative power, frequency, asymmetry, and coherence measures as classification-based univariate analysis and discriminant analysis. The study concluded with the differences of these measures between healthy and depressed patients and reported a classification accuracy of 91.3%. According to these findings, frequency-domain analysis of EEG data shows significant results for discerning between healthy and controls subjects.

Similarly, differences between healthy female controls and unmedicated depressed female patients were observed using low-resolution electromagnetic tomography (LORETA)-based analysis of cerebral activity.[6] By definition, LORETA provides localization of EEG activations based on EEG power and EEG asymmetry in different frequency bands. The differences of these activations can be studied both at the individual and group levels. In conclusion, the comparison of spectral power between groups indicated lessened activity in the right middle temporal gyrus in the depressed group.

In a different study,[7] LORETA-based analysis reported that regions of significantly increased current density in depressed patients compared to controls were generally right hemispheric, while regions of significantly decreased current density were generally frontal and left hemispheric. A within-group comparative study involving depressed subjects during two cognitive challenges reported left anterior functional hypoactivation in depression.

5.2.2 Electroencephalography Alpha Interhemispheric Asymmetry

EEG alpha interhemispheric asymmetry has been accounted as a biomarker for depression,[8,9] as many studies have reported EEG alpha interhemispheric asymmetry as significantly associated with depression. For example, left frontal hypoactivation is reported in depressed subjects only when compared with healthy controls.[10]

In study,[11] the aim was to investigate relationships among nonclinical depression/anxiety and lateralized frontal/parietotemporal activity by categorizing participants (N = 428) on the basis of both negative mood and alpha EEG. The term lateralization, in relation to brain function, means the tendency for some neural functions or cognitive processes to be specialized to one side of the brain or the other. The study investigated relationships among nonclinical depression/anxiety and lateralized frontal/parietotemporal activity. The study supported the finding that depression is associated with the frontal region of the brain but not the posterior regions.

The effect of lateralization has been observed and reported by the study.[12] Other findings show no significant differences in frontal activation between depressed and nondepressed participants.

Last,[13] a study addressed the question of whether resting anterior EEG alpha asymmetry could be considered a trait marker for depression or not. The authors concluded, "The increased variability of anterior EEG asymmetry may be a characteristic feature for depression, and, if so, this would challenge the notion that anterior EEG alpha asymmetry is a trait marker for depression."

5.2.3 Electroencephalography-Based Computer-Aided Diagnosis for Depression

This section elaborates on various ML models involved in the diagnosis of depression over the past decade. The objective is to review and elaborate on the importance of ML methods. This EEG-based research aims to investigate distinct EEG-based features (e.g., frequency bands) and their localization (e.g., the brain regions) that show association with depression. In addition, a suitable integration of feature selection methods and classification algorithms is required which could be utilized for depression detection. More recently, there has been a surge of studies utilizing ML concepts to automatically find the relationships between different EEG variables and disease outcomes.[14] The utilization of ML models allows for the discovery of nonlinear relationships as well. On the contrary, statistical methods are usually confounded with limited variable handing capability and can often be vulnerable to EEG signal nonlinearities.

5.2.3.1 Year 2018

More recently, a deep learning model has been employed for the purpose of automated screening of depression involving the convolutional neural network (CNN).[15] The authors have claimed that CNN has a profound advantage over the conventional ML model because it can automatically identify the most relevant features in EEG data. There is no need for the explicit use of the feature extracting and feature selection stages. The study has also claimed a classification accuracy of 93.54% using left hemisphere EEG data and 95.49% using right hemisphere EEG data. In conclusion, high classification results are encouraging and may be interesting for clinicians as well.

When compared with deep learning networks, convention ML models concluded with an explicit feature extraction stage, also termed as handcrafted features. In general, a ML model includes feature extraction, feature selection, classification, and validation stages. The feature extraction stage may have different features extracted from EEG data sets. For example, EEG-based functional connectivity,[16] three-channel bandwidth-duration localized wavelet filter bank,[17] discrete wavelet transformation and adaptive predictor filters,[18] spectral asymmetry index, alpha power variability and relative gamma power, nonlinear methods, Higuchi's fractal dimension, detrended fluctuation analysis and Lempel–Ziv complexity,[19] peak, variance, inclination, kurtosis, and Hjorth parameter.[14]

5.2.3.2 Year 2017

In study,[20] EEG-based features such as EEG alpha interhemispheric asymmetry and the power of different EEG frequency bands are computed and subjected to feature selection based on ROC criterion. As an implication, the most significant features are classified with a logistic regression (LR) classifier; support vector machine (SVM); and naïve Bayesian classifier. The study resulted in 98.4% accuracy, achieved with a SVM classifier.

The study,[21] aimed at designing a real-time depression monitoring system involving real-time signal filtering, artifacts removal, and power spectrum visualization. The authors claimed to develop a system that could assist clinicians during clinical decision-making. In a study,[22] a novel multiobjective particle swarm optimization (MOPSO) for depression detection is proposed. In a similar study,[23] the authors emphasized early detection and treatment of depression using the objective measures extracted from EEG data.

In study,[24] a simple method involving single channel EEG data was presented. Two different methods were compared such as the asymmetry index (SASI) and detrended fluctuation analysis (DFA). According to the results, SASI has superior discrimination ability with a classification accuracy of 76.5% compared to DFA which concluded with a classification accuracy of 70.6%. The integration of features resulted in improved accuracy, for example, the linear combination of SASI and DFA resulted in 91.2% classification accuracy.

Another study[14] utilized only three EEG channels namely Fp2, Fpz, and Fp1, making diagnosis more accessible and widespread. The feature extraction involved linear features: peak, variance, inclination, kurtosis, and the Hjorth parameter. Moreover, the nonlinear features included the Shannon, Kolmogorov, and power spectrum entropies, C0 complexity, and the correlation dimension. In addition, classification involved different classifiers such as the k-nearest neighbor (KNN), SVM, LR, decision tree (DT), and random forest (RF). The study reported the highest classification accuracy as 76.4%.

5.2.3.3 Year 2016

In the context of investigating different ML recipes, the study[25] compared combinations of different classifiers, including LR, KNN, RF, BayesNet (BN), and SVM. At the feature selection level, various feature selection methods were tested and named as the Greedy Stepwise (GSW), Genetic Search (GS), Best First (BF), Linear Forward Selection (LFS), and Rank Search (RS) based on Correlation Features Selection (CFS). Finally, the study indicated that a combination involving the feature selection method, GSW based on CFS and the classifier KNN, for the beta frequency band produced optimal classification results. The study has reported classification accuracies such as 92.00% and 98.00%. In addition, the AUCs were reported as 0.957 and 0.997. The study indicated that few EEG electrodes (FP1, FP2, F3, O2, and T3) integrated with linear features may be proficient candidates for utilization in portable systems for detecting mild depression.

For portability of EEG-based equipment, a smaller number of EEG electrodes are required. A similar system that aims for portability as well as for efficiency of EEG-based diagnostic system has been provided.[26] In this study, EEG data were collected from Fp1, Fp2, and Fpz, as these are considered closely related to emotions.

In study,[27] two different algorithms such as the neuro-fuzzy and an artificial neural network were utilized to develop an EEG-based diagnostic system for depression. Psychiatrists and scholars in Tabriz, Iran, employed a neuro-fuzzy classifier while utilizing a convenience sampling method. The study found that the neuro-fuzzy classifier was better than the artificial neural network with an achieved accuracy of 76.88%. Furthermore, the study implicated that by using EEG recorded from the electrode "F4" and the alpha peak frequency, the prediction and explanation of depression scores involving the Beck Depression Inventory (BDI) could be possible with a classification accuracy of 87%.

In study,[28] a SVM classifier with a Radial Basis Kernel Function (SVM RBF) were used. Feature extraction involved the decomposition of the EEG signal with two levels of wavelet decompositions based on discrete wavelet transform (DWT). EEG-based features such as skewness, signal energy, kurtosis, standard deviation (SD), mean, and entropy were extracted at various detailed coefficient levels of the DWT. The selection of features involved the use of the Student's *t*-test to determine the significance of differences between healthy and control subjects. Finally, the classification accuracy achieved was 88.92%. Hence, this proposed method has shown merit as a useful diagnostic tool for depression.

5.2.3.4 Year 2015

In study,[29] SVM was utilized for automatic identification of depression. The authors claimed a novel depression diagnosis index that was based on a combination of nonlinear features, including largest Lyapunov exponent, sample entropy, fractal dimension, Hurst's exponent, detrended fluctuation analysis, higher order spectra, and recurrence quantification analysis. The authors reported that the SVM classifier provided the greatest classification performance with a mean classification accuracy of about 98%, sensitivity of about 97%, and specificity of about 98.5%.

A previous study investigated the importance of nonlinear features for EEG data analysis. In study,[1] the authors claimed that nonlinear methods have significant advantages over linear methods of EEG-based feature extraction. In this study, chaos theory and nonlinear dynamic methods were widely used for extracting EEG signal features for CAD of depression.

In addition to EEG-based detection of depression, event-related potential (ERP)-based methods have shown utilization as well. For example, in a study,[30] P300 intensities and latencies were used as features to discriminate between depressed patients and healthy controls. The study utilized the SVM classifier to discriminate between depressed and healthy groups based on features such as P300 intensity values and latencies. The method resulted in a classification accuracy greater than 80%, which has promising implications for ERP-based methods as diagnostic tools for depression.

In study,[31] EEG data were decomposed using the discrete cosine transform (DCT), because the reconstruction of the EEG signal and feature formation can be performed by the DCT coefficients. Hence, the authors decomposed normal and depression patients' EEG signals into different frequency sub-bands. Moreover, feature extraction was based on the use of nonlinear methods, including sample entropy, correlation dimension, largest Lyapunov exponent, fractal dimension, Hurst exponent, and detrended fluctuation analysis. In addition, the authors performed feature ranking using *t*-test criterion. Moreover, DT, SVM, KNN), and naïve Bayes (NB) classifiers were used. Finally, the study reported the highest accuracy with the SVM classifier utilizing radial basis function (RBF), which resulted in a classification accuracy of 93.8%, a sensitivity of 92%, and a specificity of 95.8%.

In study,[32] the authors proposed a new feature selection method based on the fact that the most discriminant features could improve the classification ability of a classifier in a given setting. The authors improved on the standard Ant Colony Optimization (ACO) and termed it as improved ACO (IACO). The feature extraction involved the computation of coherence. The study utilized SVM to compute a ML model. Finally, an 80.19% overall classification accuracy was observed.

In study,[33] the possibility of a cost-effective, wearable, ubiquitous system for doctors to monitor their patients with depression was explored. The authors had used a KNN classifier to discriminate between depressed versus nondepressed individuals. The authors reported an accuracy of 99.1%.

5.2.3.5 *Brain Connectivity During Depression*

Human brain connectivity plays a vital role during normal processing of activities such as cognitive and decision-making tasks. In this context, various studies have highlighted the fact that depression exhibits abnormal brain connectivity, for example, an altered EEG functional connectivity was reported during an emotional face-word Stroop task.[34] A similar finding was reported in a different study.[35] In another study, brain connectivity was quantified using coherence involving 37 participants during the processing of facial expression stimuli.[36]

The study[37] proposed an automated EEG-based depression diagnostic and management tool based on the quantification of EEG through channel ERPs, power spectra plots, and cross-coherence. The study found an increased frontal delta (0.5–4 Hz) and alpha (8–13 Hz) power/coherence during depressed and normal/relaxed states accordingly. In addition, the study had found that devotional music (relaxed state) facilitated depression elimination. Last, the results showed statistical significance across all subjects with minimal *P*-values. In conclusion, the proposed model showed the potential to aid in early and accurate diagnosis of depression.

5.3 ELECTROENCEPHALOGRAPHY-BASED ANTIDEPRESSANT TREATMENT SELECTION

In the literature, related studies about the prediction of treatment outcomes included EEG-derived quantities such as relative and absolute power in the alpha and theta bands, QEEG theta cordance, and the antidepressant treatment response (ATR) index. Sections 5.3.1–5.3.6 provide further descriptions of these quantities. Table 5.1 illustrates studies involving electrophysiological-based scientific predictions/evidences known as EEG-based biomarkers. In addition, biomarkers show two distinct categories: (1) spontaneous/resting-state EEG predictive biomarkers and (2) ERP-based predictive biomarkers.

5.3.1 Electroencephalography Frequency Bands

EEG is a composite signal with different frequency bands, including delta, theta, alpha, beta, and gamma. However, specific bands have

TABLE 5.1 EEG Biomarkers for Antidepressant Treatment Selection

Biomarkers Class	Physiological Conditions	Derived EEG/ERP Features
EEG biomarkers	Eyes closed (EC) and open (EO)	EEG alpha band activity[38,39]
		EEG theta band activity[40,41]
		Antidepressant treatment response (ATR) index[42,43]
		EEG theta cordance[44]
		Referenced EEG (rEEG)[45]
		ML-based methods[46,47]
ERP biomarkers	Visual 3-stimulus oddball task	ERP component: P300[48,49]
		Loudness dependence auditory evoked potential (LDAEP)[50,51]

shown correlation with depression. For example, researchers have emphasized the use of EEG alpha power and EEG alpha interhemispheric asymmetry due to the relative ease of their computations. In the case of alpha power, changes in the absolute and relative powers were associated with treatment response and nonresponse. For example, Ulrich and colleagues treated 40 MDD patients with tricyclic antidepressants (TCA). They found that study participants with lateralization of alpha power at baseline were treatment responders after 4 weeks of treatment. In addition, the responders also exhibited a decrease in absolute alpha power at baseline.[52] In studies,[38,39] researchers found that early changes in the alpha band were associated with responders only. In this study, 43 depressed patients were treated with Clomipramine and Maprotiline. In a study involving 41 controls and 41 unmedicated MDD patients,[53] less alpha current source density was reported to be correlated with nonresponders when compared with both the treatment responders and the healthy controls. For the depressed patients, antidepressant treatment involved the use of SSRIs, a serotonin-norepinephrine reuptake inhibitor (SNRI), and a combined therapy with SSRI and norepinephrine-dopamine reuptake inhibitor (NDRI). On the contrary, a study found that elevated alpha power was correlated with treatment responders in 29 MDD patients treated with Imipramine for a duration of 6 weeks.[54] However, the study could not show any statistical results. Moreover, the study also reported an elevation in alpha power in combination with a reduction in theta power among treatment responders as compared with treatment nonresponders. In this study, 50 MDD patients had gone through treatment with paroxetine for a period of 6 weeks.[40]

EEG alpha interhemispheric asymmetry is defined as the alpha power differences between the left and right hemispheres. Alpha asymmetry has shown association with antidepressant treatment outcome. For example, Bruder and colleagues observed alpha asymmetry among 52 MDD patients treated with fluoxetine for 12 weeks.[55] They highlighted the presence of EEG alpha asymmetry which could differentiate between responders and nonresponders. In addition, the authors reported greater activations in the left hemisphere at baseline as compared to nonresponders. Moreover, EEG alpha asymmetry exhibited profound discriminations between female responders and female nonresponders. However, this behavior was absent in the male participants. In addition, the same medication (fluoxetine) was administered by Bruder and colleagues using 36 participants (18 controls and 18 MDD patients).[56] The authors reported 11 responders with greater baseline EEG alpha power in the occipital region as compared to nonresponders. In responders, the left hemisphere showed elevated alpha power as compared to nonresponders.

However, the generalization of results involving EEG alpha power and alpha interhemispheric asymmetry seems to be missing. There could be many reasons for this: first, because the causes of treatment efficacy are not yet clear, These could be either medication specific or individual specific; second, the studies did not provide quantitative analysis in terms of the sensitivity and specificity of the results; last, contradictory results among the studies poses a hindrance to their clinically reliability. This situation requires further investigation. Although these changes in EEG frequency bands have shown correlations, there is a need to replicate them before making a decision regarding their clinical utility.

Generally, increased EEG theta band power, also termed theta band activation, was found to be correlated with antidepressant treatment outcomes. For example, MDD patients with greater theta and lower beta band activations were declared as treatment responders when compared with treatment nonresponders.[40] Furthermore, elevated left frontal theta activity compared with the baseline was observed after electroconvulsive therapy (ECT).[41] A further follow-up, after a fourth ECT treatment, resulted in treatment response being correlated with theta activity.

In addition to theta band powers, frontal theta relative power was also investigated using 82 MDD patients treated with SSRIs for 8 weeks.[42] The results indicated significant correlation of EEG relative power at baseline and at the first week after treatment outcome. Hence, relative power was declared as a predictor of treatment outcome. Another study resulted in positive association of elevated absolute theta band power at baseline with changes of Hamilton depressive rating scale (HAM-D) scores after 8 weeks of treatment.[57] The study employed 25 MDD patients treated with SSRI, SNRI, and TCA.

However, limitations of these studies include the absence of placebo effects; therefore the ability to discriminate between medication specific and nonspecific effects is less clear and undocumented. In addition, previous studies failed to report standard results, including sensitivity, specificity, and accuracy. Hence, there is a need to validate these studies' findings.

5.3.2 Antidepressant Treatment Response Index

Numerous studies have illustrated the ATR index as a suitable predictor for antidepressant treatment response and remission involving EEG data acquisition at two times: (1) at baseline before starting the treatment and (2) at 1 week after starting treatment.[42,43] It is defined as a nonlinear fusion of three frontal QEEG parameters as described in Eq. (5.1).[42] The ATR index could have an integer value between 0 (low probability of treatment response) and 100 (high probability of treatment response):

$$\begin{aligned} ATR = \max(0,\ \min(100,\ A \times (AlphaB - AlphaA) \\ + B \times RelativeThetaPlusAlpha + C)) \end{aligned} \tag{5.1}$$

where *AlphaB* represents the absolute power in the frequency band (9–11.5 Hz) at week 1, *AlphaA* corresponds to the absolute power in the frequency band (8.5–12 Hz) at baseline, and *RelativeThetaPlusAlpha* describes relative combined theta and alpha power (3–12 Hz/2–20 Hz) at week 1. However, the seminal work did not mention any exact values of the constants *A*, *B*, and *C*.

In the literature, the ATR index has been evaluated for its accuracy. For example, the ATR index implicated into 70% correct predictions, including 82 MDD patients undergoing SSRIs treatment.[42] The study "Biomarkers for Rapid Identification of Treatment Effectiveness in Major Depression" (BRITE-MD) has replicated the ATR index in a larger sample size, that is, 220 MDD patients.[42,43] The study concluded with 74% correct treatment predictions. In this study, patients underwent treatment with escitalopram for 1 week. After that, the patients were divided into three subgroups: (1) the first group continued taking the Escitalopram; (2) the second group changed medication to Bupropion; and (3) the third group utilized both the Escitalopram and Bupropion.

The ATR index's capability to prognosticate differential response to Escitalopram, Bupropion, and their combination was investigated.[43] According to this study, an index value greater than the threshold (ATR index = 52) revealed association with the Escitalopram responders. An ATR index value smaller than the threshold (ATR index <52) was found to be associated with the Bupropion responders. However, the ATR failed to predict differential response to both Bupropion and Escitalopram. In a different study, the placebo effect for the ATR index was investigated using 23 MDD patients. Twelve of the patients were treated with Fluoxetine and 11 were treated with placebo for a duration of 8 weeks.[58] The ATR index (threshold value = 47.3) has predicted the treatment outcome with 100% sensitivity, 66.6% specificity, 75% positive predictive value, and 100% negative predictive value.

However, the studies discussed so far have limitations. First, the ATR index had utilized relatively larger sample sizes than traditional studies. This could be a reason for better performances. Second, the studies involving the ATR index had utilized different threshold values mainly selected empirically with less scientific justification. Third, due to the simplified EEG montages (smaller number of electrodes), the studies cannot be compared with traditional ones because of the different number of electrodes. Last, the ATR index formula was less clear, for example, the relative combined theta and alpha powers in the formula could not be associated with the neurobiology of depression.

5.3.3 QEEG Theta Cordance

QEEG theta cordance is a mathematically computed quantity as described by Eqs. (5.2)–(5.6).[44] In the literature, EEG absolute and relative power at different frequencies and scalp locations were utilized to compute the QEEG theta cordance as:

$$\mathrm{CORDANCE}_{(s,f)} = \pm\left(\left|a_{\mathrm{NORM}(s,f)} - 0.5\right| + \left|r_{\mathrm{NORM}(s,f)} - 0.5\right|\right) \tag{5.2}$$

$$T_s = \sum_f a_{s,f} \tag{5.3}$$

$$r_{s,f} = \frac{a_{s,f}}{T_s} \tag{5.4}$$

$$a_{\mathrm{NORM}(s,f)} = \frac{a_{s,f}}{AMAX_f} \tag{5.5}$$

$$r_{\mathrm{NORM}(s,f)} = \frac{r_{s,f}}{RMAX_f} \tag{5.6}$$

where $a_{s,f}$ is the absolute power at recording site s in the frequency band f; T_s is the total power summed up at recording site s; $r_{s,f}$ is the relative power at site s with frequency band f; AMAX_f is the maximum absolute value in frequency band f; RMAX_f is the maximum relative value in frequency band f.

In the literature, many studies have found that a decreased prefrontal QEEG theta cordance shows a correlation with treatment response.[59–64] For example, a prefrontal decrease in theta cordance 48 hours after intaking antidepressants was observed.[59,60] In addition, the decrease was found associated with treatment response when compared with treatment non-response. In these studies, response was observed 8 weeks after SSRIs and SNRIs treatment. However, in a different study, a decrease in prefrontal theta cordance after the first week of treatment (fluoxetine/venlafaxine versus placebo) was found to be associated with treatment responders only.[61] Moreover, the changes in prefrontal QEEG theta cordance at week 1 profoundly discriminated the medication responders from the medication nonresponders and the placebo responders from the nonresponders. As a result, a prediction accuracy of 72% was achieved.[65] Furthermore, accuracy was improved to 75% through treatment with other SSRIs for 8–10 weeks.[62]

The prediction ability of theta cordance was studied and resulted in similar findings. The QEEG feature such as a decrease in QEEG theta cordance observed at week one had a predicted treatment response with 88% accuracy.[63] In a separate study, 25 depressed participants were treated with venlafaxine. Treatment responders revealed lessened QEEG theta cordance values compared to nonresponders.[64,66] In addition, the study reported positive and negative predictive values as 0.7 and 0.9, respectively.

However, there are limitations associated with these studies. First, the studies involved small sample sizes that warrant replication of the findings in larger populations. Second, like the ATR index, theta cordance could not well explain the underlying neurobiology of depression.[67]

5.3.4 Referenced Electroencephalography

Referenced EEG (rEEG) uses a well-organized database. The database involves a patient's medical records regarding treatment success and failure and the associated EEG patterns. In this study, it involved observations of 1700 patients and 18,000 antidepressant treatments.[45] Recorded EEG patterns in the database allow for comparisons with a new patient's EEG patterns. Based on similar pattern matching found during previous successful treatments, the most suitable antidepressants are identified for the new patient. On the contrary, medicines with unsuccessful treatments are not selected automatically. Hence, these comparisons with the database results guided the selection or reselection of the most suitable antidepressants for individuals.

Mainly, the usefulness of rEEG-guided treatment was investigated using two groups.[45] The first group (seven subjects) was administered with rEEG-guided treatment and the second group (six subjects) was treated with normal treatment. As a result, high rates of response (six out of seven) for the rEEG-guided group were found and the second group displayed much less response (one out of six). Furthermore, the authors[68] showed improved response of rEEG-guided treatment as compared to the sequenced treatment alternative to relieve depression (STAR*D) treatment involving 114 participants. However, the studies failed to provide performance metrics in terms of sensitivity and specificity. In addition, the use of relatively smaller sample sizes requires further investigation.

5.3.5 Rostral Anterior Cingulate Cortex Activations

EEG is a modality the can record electric potentials from the human scalp; however it cannot localize the actual brain sources responsible for scalp recordings. Brain source localization (BSL) techniques such as

LORETA and its variants, including standardized LORETA (sLORETA), standardized shrinking LORETA-FOCUSS (ssLOFO), and more, are used to localize brain regions. Studies based on BSL have identified activations of brain areas that are associated with treatment outcomes as either treatment responders or nonresponders. For example, in studies,[69,70] 18 MDD patients were treated with Nortriptyline, Citalopram and Reboxitine for 16 weeks. Both studies resulted in increased pretreatment resting delta activity in the rostral anterior cingulate cortex (rACC). The activity was associated with treatment response. In another study, higher theta activity in the rACC and the orbitofrontal cortex was found to be associated with response to medication.[71]

Table 5.2 summarized these studies. EEG biomarkers have shown correlations with treatment outcomes for depression. Neurobiology in support of this association has been less clear and is described in the respective summary tables.

TABLE 5.2 EEG Studies Involving Treatment Selection for Depression

EEG Biomarker	Brain Dynamics	Critical Analysis and Limitations
Alpha power and asymmetry (alpha power and alpha asymmetry lateralization of alpha power)[38–40,52–58]	Treatment responders showed decreased lateralization and alpha power. Moreover, less current source density was found to be correlated with treatment nonresponse. Increased alpha power was correlated with treatment response only.	According to the literature, EEG power measurements are considered reliable. In addition, changes associated with alpha asymmetry could discriminate between treatment responders and nonresponders.
Theta band activations (theta band powers)[40–42,57]	The studies reported an association between treatment response and increased theta activation.	Decreased alpha activity was associated with treatment response.[42,54] In addition, increased theta band activity has also been reported.[40]
rACC activations (EEG power computations and their LORETA-based source localization)	Activation in rACC was associated with treatment response only.	rACC activations were found to be consistent during depression. However, more effort is needed to justify the empirically adjusted thresholds.

(*Continued*)

TABLE 5.2 (Continued)

EEG Biomarker	Brain Dynamics	Critical Analysis and Limitations
QEEG theta cordance (absolute and relative power in theta bands)[44,59–64]	Decreased QEEG theta cordance was found to be correlated with treatment response.	Decreased QEEG theta cordance was declared as a predictive biomarker. However, associated low specificity values restricted their application for clinical purposes.
ATR index (relative theta and alpha power)[42,43,58]	The high classification accuracy implicated better predictions involving the ATR index.In addition, the ATR index can be a differential indicator for different kinds of antidepressants.	The ATR index was utilized with empirically adjusted threshold values. These adjustments involved experimental observations and may need further validation for clinical applications.
Referenced EEG (rEEG)[45,68]	rEEG-guided treatment showed improvement over the STAR*D study. Hence, it implicated improvement of evidence-based therapy over the sequential treatment strategy.	rEEG had performed better than STAR*D. However, the database was specific to one geographical area. For generalization, patterns from other geographical areas should be included.

5.3.6 Machine Learning Methods for Treatment Selection

ML techniques involving antidepressant selection performed an automatic identification of the most suitable patterns specific to either the disease conditions or the healthy controls. In general, ML methods include subprocesses such as feature extraction, feature selection, classification, and validation. More specifically, feature extraction has fundamental importance that aims to derive information after analyzing either EEG or ERP data.[72,73] Moreover, the purpose of ML methods is to introduce automation during the diagnosis and treatment selection for MDD patients. For example, a ML method was developed named "Psychiatrist in the Machine."[74] The system could assist MDD patients in getting suitable treatment according to their EEG data. Furthermore, a different study has shown improved results while performing treatment outcome prediction for MDD patients. The method has provided high classification results (sensitivity = 94.9%, specificity = 80.9%, accuracy = 87.9%).[75] In addition, a similar study was performed for schizophrenia patients that concluded with a classification accuracy of 85%.[76] Table 5.3 summarizes further details on the latest ML methods for depression.

TABLE 5.3 ML Methods for Treatment Selection During Depression

Brain Dynamics	Year	Sample Size	Classification Accuracy (%)
Using pretreatment EEG data to predict responses to SSRI treatment for MDD.[77]	2010	22	85
The authors investigated a pilot study fusing ML methodologies with EEG to predict symptomatic responses to clozapine therapy.[76]	2012	23	85
Various EEG features were computed, including EEG-based coherences.[75]	2013	22	87.9
Wavelet-based decomposition of EEG data was performed and the coefficients were utilized as EEG biomarkers.[78]	2017	33	87.5

5.4 EVENT-RELATED POTENTIAL-BASED ANTIDEPRESSANT TREATMENT SELECTION

ERPs involve the responses of the human brain to sensory stimuli (visual and auditory). Various research studies[48,51,79–81] have observed P300 and the loudness-dependence of the auditory evoked potential (LDAEP) as predictive biomarkers for depression.

5.4.1 P200 and P300

P200 and P300 are ERP components based on the computations of grand averages of ERP data associated with certain stimuli types. For example, P300 could be computed for either targets or nontargets. Morphological quantities involving P300 have been associated with treatment outcome for depression. The quantities may be either P300 amplitudes or P300 latencies. Normally, peak P300 intensity occurs between 300 and 700 ms with a prominent amplitude after the occurrence of stimuli. Abnormalities associated with P300 could include low P300 intensity and delayed P300 peak, termed as less amplitudes and more latencies accordingly. Since P300 has association with cognitive abilities, in depressed participants, any abnormalities in P300 could implicate cognitive dysfunctions and may be associated with depression. In the literature, studies involving P300 show correlation with treatment outcome involving depression. For example, P300 peak amplitudes at the occipital electrodes were found to be associated with treatment response during a dichotic listening task.[79] In study,[48] 49 MDD patients and 22 controls were recruited. The study reported that depressed patients with larger P300 latency at baseline failed to remit after 6 weeks. On the other hand, remitted patients along with healthy controls exhibited normal latencies. In addition, 23 out of 24 remitted MDD patients were identified by P300 with a specificity of 65.9% and a sensitivity of 95.8%.

In an observational study,[80] treatment response at baseline was found to be associated with higher P200 slopes compared with nonresponse. Moreover, in a 12-week study based on 28 MDD patients treated with sertraline, longer latencies were reported in nonrespondents (10) as compared with both respondents (18) and controls (28).[49] However, no correlations were found between P300 intensities and treatment response. In an early study,[80] 17 MDD patients (11 respondents and 6 nonrespondents) were recruited for 4–8 weeks and treated with Fluoxetine, Bupropion, or Desipramine. As a result, larger P200 slopes at baseline were reported in respondents only when compared with nonrespondents.

5.4.2 Loudness Dependence Auditory Evoked Potential

Studies involving loudness dependence auditory evoked potential (LDAEP) have reported association with treatment outcome. For example, higher LDAEP slopes were observed to be associated with treatment respondents only.[50] In a different study, lower LDAEP slopes at baseline were correlated with treatment response to Reboxetine.[51] Furthermore, higher values of LDAEP in respondents were observed as compared to nonrespondents.[81]

LDAEP is capable of being a differential marker for two types of medications with different mechanisms-of-action (MOA): serotonergic and nonserotonergic. In Ref. [82], 20 of the depressed patients were administered with Citalopram (an SSRI) and another15 depressed patients were treated with Reboxetine (an NRI). The former group of patients showed higher LDAEP slopes at baseline in contrast to the Reboxetine respondents. Similar results were replicated by.[83] In Ref. [84], low LDAEP at baseline was found in an MDD patient (only one patient) with severe side effects to SSRI treatment. Moreover, medication with a different neurochemical profile, named Tianeptine, showed usefulness instead.

However, P300-based studies had limitations: the studies had utilized distinctive features, for example, P300 amplitude or latency. Due to these differences, it is impossible to compare them with each other. Hence, the generalization of results was not possible. Moreover, the studies failed to report sensitivity and specificity values. LDAEP showed a link with serotonergic activity, but it was not fully

TABLE 5.4 ERP Studies for Treatment Selection During Depression

ERP Biomarker	Key Findings	Limitations	Neurobiological Description
P300 (amplitude and latencies), P200 (slopes)[49,79,80]	Treatment responders have shown correlation with higher P300 amplitude as well as higher P200 slopes. Nonresponders have shown larger P300 latencies.	Comparison within studies was not possible because the studies discussed different ERP parameters. In addition, the studies did not report standard results such as the classification of sensitivities and specificities.	The exact neurobiological basis of P300 is unclear. However, a study associated P300 latency with response to 5-HT agonist Flesinoxan.[85] This finding is significant because the 5-HT receptor plays an important role in the therapeutic actions of antidepressants.
LDAEPEPR component (N100/P200) changes with increasing loudness of auditory stimulus[82,84]	Responders have shown higher LDAEP slopes. LDAEP is a differential predictor for two distinct drugs.	Studies found thresholds mainly based on empirical findings, which needs further validations.	LDEAP is hypothesized to link with 5-HT activity.[86] Higher levels of 5-HT were found to be related to ERP suppression involving responsiveness to auditory tone intensity. On the other hand, low 5-HT levels facilitate ERP.

demonstrated that LDAEP might differentiate between the response to serotonergic and nonserotonergic antidepressants. The clinical utility of LDAEP needs further validation. Table 5.4 shows a summary of the studies.

5.5 INTEGRATING NEUROBIOLOGICAL AND ELECTROPHYSIOLOGICAL DATA

The literature has evidenced that integrating various EEG parameters could enhance the diagnosis and treatment selection for depression. For example, fusing alpha spectral power (sensitivity 73% and specificity 58%)

and alpha interhemispheric asymmetry (sensitivity 64% and specificity 71%) gives an overall sensitivity of 83% and specificity of 68%.[56] In ddition, integrating theta activity localized to the rACC with LORETA and the LDAEP technique have shown positive improvements.[87] More specifically, the integration of the two techniques has proven to profoundly differentiate both treatment responders and nonresponders.

Furthermore, the inception of powerful techniques based on ML has resulted in powerful knowledge-based systems. EEG features such as EEG alpha power, alpha asymmetry, ACC theta power, LDAEP, and connectivity measures between different brain regions may all have the potential to diagnose and predict treatment efficacy for depression. One possible integration technique is to concatenate different EEG-derived features into a common data matrix. The matrix would be used as the input data for pattern recognition techniques. After proper training of the ML model, automatic identification of disease-specific patterns could be possible. Similarly, the identification of EEG patterns specific to either treatment responders or nonresponders could be possible as well.

Many studies have posited that features based on EEG and ERP data were capable for diagnosing and predicting treatment outcome for depression. Conventional methods of diagnosing MDD involve symptom assessments based on clinical questionnaires. These methods follow the procedure as described in the Diagnostic and Statistical Manual of Depression-V (DSM-V), which is considered as standard. However, some symptoms may overlap with other mental illnesses such as bipolar depression and schizophrenia. Hence, unipolar depression could be misdiagnosed, for example, as bipolar depression.

More than 20 different antidepressants are available under the category of SSRIs. However, their selection for a patient is subjective and based on professional expertise. Improvement in the selection process is possible based on objective evidence generated by neuroimaging modalities such as EEG and ERP. Hence, both for diagnosis and treatment selection, objective measures based on EEG and ERP modalities are required. These modalities may provide additional evidence to perform more accurate and valid diagnosis and assessment of antidepressants' efficacies.

In this context, features such as power computations of various EEG frequency bands and the computation of EEG alpha asymmetries have provided enough evidence to discriminate MDD patients from controls. In addition to EEG, ERP features such as P300 intensities and latencies show promise. A further improvement could be an integration of EEG and ERP features. This integration could possibly result in robust ML models.

5.6 SUMMARY

In this chapter, various EEG and ERP studies have shown promise in discriminating and predicting the treatment outcome for depression. During the literature review, the clinical utility of these methods has revealed a lack of clarity due to certain limitations: the study samples are heterogeneous because MDD is heterogeneous in nature. For the sake of generalization, more studies should involve homogenous study samples. Moreover, the studies did not present their results in standard metrics such as accuracy, sensitivity, and specificities. Hence, a common conclusion was impossible, as comparison among studies was not possible. In general, EEG-based methods have been less successful in explaining the whole variance of depression.

References

1. Acharya UR, Sudarshan VK, Adeli H, Santhosh J, Koh JE, Adeli A. Computer-aided diagnosis of depression using EEG signals. *Eur Neurol*. 2015;73:329–336.
2. Sun Y, Li Y, Zhu Y, Chen X, Tong S. Electroencephalographic differences between depressed and control subjects: an aspect of interdependence analysis. *Brain Res Bull*. 2008;76:559–564.
3. Wan Z, He Q, Zhou H, Yang J, Yan J, Zhong N. A quantitative analysis method for objectively assessing the depression mood status based on portable EEG and self-rating scale. In: Zeng Y, et al., eds. Brain Informatics, *BI 2017. Lecture Notes in Computer Science*, vol. 10654. Cham: Springer; 2017.
4. Moore P, Van Pham H. On context-aware evidence-based data driven development of diagnostic scales for depression. In: Barolli L, Terzo O, eds. Complex, Intelligent, and Software Intensive Systems. *CISIS 2017. Advances in Intelligent Systems and Computing*, vol. 611. Cham: Springer; 2018.
5. Knott V, Mahoney C, Kennedy S, Evans K. EEG power, frequency, asymmetry and coherence in male depression. *Psychiatry Res*. 2001;106:123–140.
6. Lubar JF, Congedo M, Askew JH. Low-resolution electromagnetic tomography (LORETA) of cerebral activity in chronic depressive disorder. *Int J Psychophysiol*. 2003;49:175–185.
7. Flor-Henry P, Lind JC, Koles ZJ. A source-imaging (low-resolution electromagnetic tomography) study of the EEGs from unmedicated males with depression. *Psychiatry Res Neuroimaging*. 2004;130:191–207.
8. Dharmadhikari A, Tandle A, Jaiswal S, Sawant V, Vahia V, Jog N. Frontal theta asymmetry as a biomarker of depression. *East Asian Arch Psychiatry*. 2018;28:17.
9. Allen JJ, Urry HL, Hitt SK, Coan JA. The stability of resting frontal electroencephalographic asymmetry in depression. *Psychophysiology*. 2004;41:269–280.
10. Henriques JB, Davidson RJ. Left frontal hypoactivation in depression. *J Abnorm Psychol*. 1991;100:535.
11. Mathersul D, Williams LM, Hopkinson PJ, Kemp AH. Investigating models of affect: relationships among EEG alpha asymmetry, depression, and anxiety. *Emotion*. 2008;8:560.
12. Reid SA, Duke LM, Allen JJ. Resting frontal electroencephalographic asymmetry in depression: inconsistencies suggest the need to identify mediating factors. *Psychophysiology*. 1998;35:389–404.

13. Debener S, Beauducel A, Nessler D, Brocke B, Heilemann H, Kayser J. Is resting anterior EEG alpha asymmetry a trait marker for depression? *Neuropsychobiology*. 2000;41:31–37.
14. Cai H, Chen Y, Han J, Zhang X, Hu B. Study on feature selection methods for depression detection using three-electrode EEG data. *Interdiscip Sci*. 2018;10:1–8.
15. Acharya UR, Oh SL, Hagiwara Y, Tan JH, Adeli H, Subha D. Automated EEG-based screening of depression using deep convolutional neural network. *Comput Methods Program Biomed*. 2018;161:103–113.
16. Mumtaz W, Ali SSA, Yasin MAM, Malik AS. A machine learning framework involving EEG-based functional connectivity to diagnose major depressive disorder (MDD). *Med Biol Eng Comput*. 2018;56:233–246.
17. Sharma M, Achuth P, Deb D, Puthankattil SD, Acharya UR. An automated diagnosis of depression using three-channel bandwidth-duration localized wavelet filter bank with EEG signals. *Cogn Syst Res*. 2018;52:508–520.
18. Cai H, Han J, Chen Y, et al. A pervasive approach to EEG-based depression detection. *Complexity*. 2018;2018:1–13.
19. Bachmann M, Päeske L, Kalev K, et al. Methods for classifying depression in single channel EEG using linear and nonlinear signal analysis. *Comput Methods Programs Biomed*. 2018;155:11–17.
20. Mumtaz W, Xia L, Ali SSA, Yasin MAM, Hussain M, Malik AS. Electroencephalogram (EEG)-based computer-aided technique to diagnose major depressive disorder (MDD). *Biomed Signal Process Control*. 2017;31:108–115.
21. Zhao S, et al. Wearable EEG-based real-time system for depression monitoring. In: Zeng Y, et al., eds. Brain Informatics. *BI 2017. Lecture Notes in Computer Science*, vol. 10654. Cham: Springer; 2017.
22. Guo Y, Zhang H, Pang C. EEG-based mild depression detection using multi-objective particle swarm optimization. In: *Control And Decision Conference (CCDC), 2017 29th Chinese*. Changqing, China; 2017:4980–4984.
23. Shen J, Zhao S, Yao Y, Wang Y, Feng L. A novel depression detection method based on pervasive EEG and EEG splitting criterion. In: *Bioinformatics and Biomedicine (BIBM), 2017 IEEE International Conference on*. Kansas City, MO, USA; 2017:1879–1886.
24. Bachmann M, Lass J, Hinrikus H. Single channel EEG analysis for detection of depression. *Biomed Signal Process Control*. 2017;31:391–397.
25. Li X, Hu B, Sun S, Cai H. EEG-based mild depressive detection using feature selection methods and classifiers. *Comput Methods Program Biomed*. 2016;136:151–161.
26. Cai H, Sha X, Han X, Wei S, Hu B. Pervasive EEG diagnosis of depression using deep belief network with three-electrodes EEG collector. In: *Bioinformatics and Biomedicine (BIBM), 2016 IEEE International Conference on*. Shenzhen, China; 2016:1239–1246.
27. Mohammadzadeh B, Khodabandelu M, Lotfizadeh M. Comparing diagnosis of depression in depressed patients by EEG, based on two algorithms: Artificial Nerve Networks and Neuro-Fuzy Networks. *Int J Epidemiol Res*. 2016;3:246–258.
28. Bairy MG, Niranjan U, Puthankattil SD. Automated classification of depression EEG signals using wavelet entropies and energies. *J Mech Med Biol*. 2015;16:1–13.
29. Acharya UR, Sudarshan VK, Adeli H, et al. A novel depression diagnosis index using nonlinear features in EEG signals. *Eur Neurol*. 2015;74:79–83.
30. Mumtaz W, Malik AS, Ali SSA, Yasin MAM. P300 intensities and latencies for major depressive disorder detection. In: *Signal and Image Processing Applications (ICSIPA), 2015 IEEE International Conference on*. Kuala Lumpur, Malaysia:542–545.
31. Bairy GM, Bhat S, Eugene LWJ, Niranjan U, Puthankattil SD, Joseph PK. Automated classification of depression electroencephalographic signals using discrete cosine transform and nonlinear dynamics. *J Med Imag Health Inform*. 2015;5:635–640.
32. Erguzel TT, Tas C, Cebi M. A wrapper-based approach for feature selection and classification of major depressive disorder–bipolar disorders. *Comput Biol Med*. 2015;64:127–137.

33. Li X, Hu B, Shen J, Xu T, Retcliffe M. Mild depression detection of college students: an EEG-based solution with free viewing tasks. *J Med Syst*. 2015;39:187.
34. Guo Z, Wu X, Liu J, Yao L, Hu B. Altered Electroencephalography Functional Connectivity in Depression During the Emotional Face-Word Stroop Task. *J Neural Eng*. 2018;15:056014.
35. Guo Z, Long H, Yao L, Wu X, Cai H. Abnormal EEG-based functional connectivity under a face-word stroop task in depression. In: *Bioinformatics and Biomedicine (BIBM), 2017 IEEE International Conference on*. Kansas City, MO, USA; 2017:946–951.
36. Li X, Jing Z, Hu B, Sun S, An EEG-based study on coherence and brain networks in mild depression cognitive process. In: *Bioinformatics and Biomedicine (BIBM), 2016 IEEE International Conference on*. Shenzhen, China; 2016:1275–1282.
37. Mahajan R, Bansal D. Depression diagnosis and management using EEG-based affective brain mapping in real time. *Int J Biomed Eng Technol*. 2015;18:115–138.
38. Ulrich G, Haug H, Fahndrich E. Acute vs. chronic EEG effects in maprotiline- and in clomipramine-treated depressive inpatients and the prediction of therapeutic outcome. *J Affect Disord*. 1994;32:213–217.
39. Ulrich G, Haug H, Stieglitz R, Fahndrich E. EEG characteristics of clinically defined on-drug-responders and nonresponders—a comparison clomipramine vs.maprotiline. *Pharmacopsychiatry*. 1988;21:367–368.
40. Knott V, Mahoney C, Kennedy S, Evans K. Pre-treatment EEG and its relationship to depression severity and paroxetine treatment outcome. *Pharmacopsychiatry*. 2000;33:201–205.
41. Heikman P, Salmelin R, Mäkelä JP, Hari R, Katila H, Kuoppasalmi K. Relation between frontal 3–7 Hz MEG activity and the efficacy of ECT in major depression. *J ECT*. 2001;17:136–140.
42. Leuchter AF, Cook IA, Marangell LB, et al. Comparative effectiveness of biomarkers and clinical indicators for predicting outcomes of SSRI treatment in major depressive disorder: results of the BRITE-MD study. *Psychiatry Res*. 2009;169:124–131.
43. Leuchter A, Cook I, Gilmer W. Effectiveness of a quantitative electroencephalographic biomarker for predicting differential response or remission with escitalopram and bupropion in major depressive disorder. *Psychiatry Res*. 2009;169:132–138.
44. Leuchter AF, Cook IA, Lufkin RB, et al. Cordance: a new method for the assessment of cerebral perfusion and metabolism using quantitative encephalography. *NeuroImage*. 1994;1:208–219.
45. Suffin SC, Emory WH, Gutierrez N, Arora GS, Schiller MJ, Kling A. A QEEG database method for predicting pharmacotherapeutic outcome in refractory major depressive disorders. *J Am Phys Surg*. 2007;12.
46. Khodayari-Rostamabad A, Reilly HP, Hasey GM, Bruin HD, MacCrimmon DJ. A machine learning approach using EEG data to predict response to SSRI treatment for major depressive disorder. *Clin Neurophysiol*. 2013;xxx:1–11.
47. Khodayari-Rostamabad A, Hasey GM, MacCrimmon DJ, Reilly JP, Bruin HD. A pilot study to determine whether machine learning methodologies using pre-treatment electroencephalography can predict the symptomatic response to clozapine therapy. *Clin Neurophysiol*. 2010;121:1998–2006.
48. Kalayam B, Alexopoulos GS. Prefrontal dysfunction and treatment response in geriatric depression. *Arch Gen Psychiatry*. 2011;56:713–718.
49. Isıntas M, Ak M, Erdem M, Oz O, Ozgen F. Event-related potentials in major depressive disorder: the relationship between P300 and treatment response. *Turk Psikiyatri Derg*. 2012;23:33–39.
50. Lee TW, Yu YWY, Chen T-J, Tsai S-J. Loudness dependence of the auditory evoked potential and response to antidepressants in Chinese patients with major depression. *Rev Psychiatry Neurosci*. 2004;30:202–205.
51. Linka T, Müller B, Bender S, Sartory G, Gastpar M. The intensity dependence of auditory evoked ERP components predicts responsiveness to reboxetine treatment in major depression. *Pharmacopsychiatry*. 2005;38:139–143.

52. Ulrich G, Renfordt E, Zeller G, Frick K. Interrelation between changes in the EEG and psychopathology under pharmacotherapy for endogenous depression. A contribution to the predictor question. *Pharmacopsychiatry*. 1984;17:178–183.
53. Tenke CE, Keyser J, Manna CG, et al. Current source density mesures of EEG alpha predict antidepressant treatment response. *Biol Pschiatry*. 2011;70:388–394.
54. Knott V, Telner J, Lapierre Y, Browne M, Horne E. Quantitative EEG in the prediction of antidepressant response to imipramine. *J Affect Disord*. 1996;39:175–184.
55. Bruder GE, Stewart JW, Tenke CE, et al. Electroencephalographic and perceptual asymmetry differences between responders and nonresponders to an SSRI antidepressant. *Biol Psychiatry*. 2001;49:416–425.
56. Bruder GE, Sedoruk JP, Stewart JW, McGrath PJ, Quitkin FM, Tenke CE. Electroencephalographic alpha measures predict therapeutic response to a selective serotonin reuptake inhibitor antidepressant: pre- and post-treatment findings. *Biol Psychiatry*. 2008;63:1171–1177.
57. Spronk D, Arns M, Barnett KJ, Cooper NJ, Gordon E. An investigation of EEG, genetic and cognitive markers of treatment response to antidepressant medication in patients with major depressive disorder: a pilot study. *J Affect Disord*. 2011;128:41–48.
58. Hunter A, Cook I, Greenwald S, Tran M, Miyamoto K, Leuchter A. The antidepressant treatment response index and treatment outcomes in a placebo-controlled trial of fluoxetine. *J Clin Neurophysiol*. 2011;28:478–482.
59. Leuchter A, Cook I, Uijtdehaage S, et al. Brain structure and function and the outcomes of treatment for depression. *J Clin Psychiatry*. 1997;58:22–31.
60. Cook I, Leuchter A. Prefrontal changes and treatment response prediction in depression. *Semin Clin Neuropsychiatry*. 2001;6:113–120.
61. Cook I, Leuchter A, Morgan M, et al. Early changes in prefrontal activity characterize clinical responders to antidepressants. *Neuropsychopharmacology*. 2002;27:120–131.
62. Cook IA, Leuchter AF, Morgan ML, Stubbeman W, Siegman B, Abrams M. Changes in prefrontal activity characterize clinical response in SSRI nonresponders: a pilot study. *J Psychiatry Res*. 2005;39:461–466.
63. Bares M, Brunovsky M, Kopecek M, et al. Changes in QEEG prefrontal cordance as a predictor of response to antidepressants in patients with treatment resistant depressive disorder: a pilot study. *J Psychiatry Res*. 2007;41:319–325.
64. Bares M, Brunovsky M, Kopecek M, et al. Early reduction in prefrontal theta QEEG cordance value predicts response to venlafaxine treatment in patients with resistant depressive disorder. *Eur Psychiatry*. 2008;23:350–355.
65. Leuchter AF, Cook IA, Witte EA, Morgan M, Abrams M. Changes in brain function of depressed subjects during treatment with placebo. *Am J Psychiatry*. 2002;159:122–129.
66. Bares M, Brunovsky M, Novak T, et al. The change of prefrontal QEEG theta cordance as a predictor of response to bupropion treatment in patients who had failed to respond to previous antidepressant treatments. *Eur Neuropsychopharmacol*. 2010;20:459–466.
67. Baskaran A, Milev R, McIntyre R. The neurobiology of the EEG biomarker as a predictor of treatment response in depression. *Neuropharmacology*. 2012;63:507–513.
68. DeBattist C, Kinrys G, Hoffman D, et al. The use of referenced-EEG (rEEG) in assisting medication selection for the treatment of depression. *J Psychiatry Res*. 2011;45:64–75.
69. Pizzagalli D, Marqui RDP, Nitschke JB, et al. Anterior cingulate activity as a predictor of degree of treatment response in major sepression: evidence from brain electrical tomography analysis. *Am J Psychiatry*. 2001;158:405–415.
70. Mulert C, Juckel G, Brunnmeier M, et al. Rostral anterior cingulate cortex activity in the theta band predicts response to antidepressive medication. *Clin EEG Neurosci*. 2007;38:78–81.

71. Korb AS, Hunter AM, Cook IA, Leuchter AF. Rostral anterior cingulate cortex theta current density and response to antidepressants and placebo in major depression. *Clin Neurophysiol*. 2009;120:1313–1319.
72. Lotte F, Congedo M, Lecuyer A, Lamarch F, Arnaldi B. A review of classication algorithms for EEG-based brain-computer interfaces. *J Neural Eng*. 2007;4:1–24.
73. Yu Z-E, Kuo C-C, Chou C-H, Yen C-T, Chang F. A machine learning approach to classify vigilance states in rats. *Expert Syst Appl*. 2011;38:10153–10160.
74. Moore SK. The psychiatrist in the machine [Update]. *IEEE Spectrum*. 2011;48(3):11–12.
75. Khodayari-Rostamabad A, Reilly JP, Hasey GM, de Bruin H, MacCrimmon DJ. A machine learning approach using EEG data to predict response to SSRI treatment for major depressive disorder. *Clin Neurophysiol*. 2013;124:1975–1985.
76. Khodayari-Rostamabad A, Hasey GM, MacCrimmon DJ, Reilly JP, de Bruin H. A pilot study to determine whether machine learning methodologies using pre-treatment electroencephalography can predict the symptomatic response to clozapine therapy. *Clin Neurophysiol*. 2010;121:1998–2006.
77. Khodayari-Rostamabad A, Reilly JP, Hasey G, MacCrimmon D. Using pre-treatment EEG data to predict response to SSRI treatment for MDD. In: *Engineering in Medicine and Biology Society (EMBC), 2010 Annual International Conference of the IEEE*; Buenos Aires, Argentina, 2010:6103–6106.
78. Mumtaz W, Xia L, Yasin MAM, Ali SSA, Malik AS. A wavelet-based technique to predict treatment outcome for major depressive disorder. *PLoS One*. 2017;12:e0171409.
79. Bruder GE, Tenke CE, Stewart JW, et al. Brain event-related potentials to complex tones in depressed patients: relations to perceptual, asymmetry and clinical features. *Psychophysiology*. 1995;32:373–381.
80. Paige R, Fitzpatrick D, Kline J, Balogh S, Hendricks S. Event-related potential amplitude/intensity slopes predict response to antidepressants. *Neuropsychobiology*. 1994;30:197–201.
81. Gallinat J, Bottlender R, Juckel G, et al. The loudness dependency of the auditory evoked N1/P2-component as a predictor of the acute SSRI response in depression. *Psychopharmacology (Berl)*. 2000;148:139–143.
82. Juckel G, Pogarell O, Augustin H, et al. Differential prediction of first clinical response to serotonergic and noradrenergic antidepressants using the loudness dependence of auditory evoked potentials in patients with major depressive disorder. *J Clin Psychiatry*. 2007;68:1206–1212.
83. Linka T, Sartory G, Wiltfang J, Müller B. Treatment effects of serotonergic and noradrenergic antidepressants on the intensity dependence of auditory ERP components in major depression. *Neurosci Lett*. 2009;463:26–30.
84. Park Y-M, Lee S-H, Park EJ. Usefulness of LDAEP to predict tolerability to SSRIs in major depressive disorder: a case report. *Psychiatry Investig*. 2012;9:80–82.
85. Hansenne M, Ansseau M. P300 event-related potential and serotonin-1A activity in depression. *Eur Psychiatry*. 1999;14:143–147.
86. Hegerl U, Juckel G. Intensity dependence of auditory evoked potentials as an indicator of central serotonergic neurotransmission: a new hypothesis. *Biol Psychiatry*. 1993;33:173–187.
87. Mulert C, Juckel G, Brunnmeier M, et al. Prediction of treatment response in major depression: integration of concepts. *J Affect Disord*. 2007;98:215–225.

CHAPTER

6

Neural Circuits and Electroencephalography-Based Neurobiology for Depression

6.1 INTRODUCTION

Depression is a debilitating and recurrent mental illness and plays a key role in functional disability. It constitutes the number one source of "years lived with disability" (YLDs).[1] However, until recently, its pathogenesis was less explored and remained unclear. There could be two possible reasons for this: (1) major depressive disorder (MDD) has not been associated with whole brain pathology and (2) there is an absence of clear animal models for specific mood episodes. However, the development of neuroimaging technologies has helped during in vivo characterization of neurochemistry, physiology, and anatomy in human subjects with mental disorders. In addition, these technologies have enhanced the advancement toward elucidation of the pathophysiology of various mental disorders. However, the interpretation of these abnormalities is mainly based on collateral descriptions of anatomical networks that explain emotional behavior.

This chapter has two main sections: the first part discusses the neuroanatomy of neural circuits implicated during depression based on a synthesis of findings from nonhuman primates and human subjects referred from clinical studies. The techniques are mainly based on postmortem, lesion analysis, and neuroimaging methodologies. The pathophysiology of depression involves a dysfunction in the network, including the medial prefrontal cortex (mPFC) and anatomically related striatal, limbic, basal, and thalamic forebrain structures.

Section 6.2 elaborates on the neurobiology of EEG-based predictive biomarkers for depression It refers to the underlying brain mechanisms that could be associated with the EEG-based biomarker for depression.

EEG-Based Experiment Design for Major Depressive Disorder
DOI: https://doi.org/10.1016/B978-0-12-817420-3.00006-0

EEG-based biomarkers such as the QEEG theta cordance, alpha power, alpha asymmetry, and the antidepressant treatment response (ATR) index were explained in Chapter 5, Using Electroencephalography for Diagnosing and Treating Depression.

6.2 NEURAL CIRCUITRY IMPLICATED DURING DEPRESSION

This section explains anatomical structural changes which could cause or become effected during depression. In addition, abnormal connections between brain regions that occur because of depression are discussed in detail. In general, connections between cortical areas, mainly the frontal cortical areas such as the anterior cingulate cortex (ACC) and midbrain regions such as the amygdala, hypothalamus and related brain regions, have been reported in many studies and mentioned in Sections 6.2.1–6.2.6.

6.2.1 Clinical Phenomenology of Depression

The clinical phenomenology of depression implies the identification of risk factors that could cause it. The clinical phenomenology of major depression has identified the brain systems involved in reward processing, anxiety, mood, fear, motivation, attention, and stress responses.[1] In addition, either anhedonia or a depressed mood could become a cause to establish the diagnosis of a major depressive episode.[1] The term anhedonia refers to the inability to experience pleasure.[2] According to the *Diagnostic and Statistical Manual V* (*DSM-V*),[1] anhedonia is defined as diminishing interest or pleasure in response to reward-related stimuli. Anhedonia is common among depressed individuals, it has been reported that between one-third and half of MDD patients experience clinically notable anhedonia.[2]

Many studies have supported the fact that the depressed subjects show insensitivity to reward stimulus. For example, studies based on reinforcement paradigms to investigate anhedonia during depression reported that study participants with depressive symptoms failed to achieve a response bias toward reward stimuli.[3–5]

As with anhedonia, neuropsychological and cognitive impairments are characteristics of major depression. In the *DSM-V*, it is reflected as "an inability to think or concentrate."[1] However, conflicting results can be found as some studies have reported wide-ranging deficits such as impairments in early information processing, attention, memory, and executive functions. On the contrary, other studies could not report such findings.[6,7] These contradictions may be due to heterogeneity with the study participants and medication status. For example, impairment of spatial memory and delayed matching is found in unmedicated MDD patients as compared to medicated MDD patients.[8,9]

A study concluded that depressed subjects exhibit a mood congruent processing bias, meaning a tendency toward biased stimulus processing, particularly to negative information as compared to positive or neutral information.[10–12] Similar findings were reported in other studies, for example, in studies on memory or attention, biasing stimulus information toward sad information is confirmed by enhanced recall for negatively versus positively balanced information on memory tests.[12,13] In the stroop test, depressed subjects show greater interference from depressed-related negative words versus happy or neutral words.[14,15] Further findings were presented on speedy responses to sad versus happy words during affective attention shifting tasks.[16] During face dot-probe tasks, preferential attentiveness to faces with sad versus neutral expressions was reported.[17,18] Furthermore, during probabilistic reward tasks, depressed subjects showed oversensitivity to negative feedback.[7,8]

6.2.2 Neural Substrates of Mood Disorders

Imaging, histopathological, and deep brain stimulation studies in humans have revealed that related medial prefrontal cortical areas and amygdala are involved during mood disorders.[19] As shown in Fig. 6.1, experimental anatomical evidence from animals supported the findings reported in human studies. In Fig. 6.1, according to animal studies, the connectional network involves several areas in the medial mPFC and the amygdala.[19,20]

As shown in Figs. 6.2 and 6.3,[9, 22–24] brain regions such as the amygdala, hippocampus, and the "limbic" system are considered as the central parts of the emotional brain. In addition, studies based on axonal transport have identified it as a system that links the mPFC and a few related cortical areas to the amygdala, the pallidum and ventral striatum, the hypothalamus, the medial thalamus, and the periaqueductal gray and other parts of the brainstem. In addition to animal studies, evidence from human studies confirm a similar system, including functional and structural imaging and the analysis of lesions and histological materials. Figs. 6.2 and 6.3 focus on various aspects of brain structure such as the Fig. 6.2 has elaborated the prefrontal cortex. On the other hand, Fig. 6.3 focuses more on the midbrain connections.

In Fig. 6.2, brain networks are shown that connect the medial network and amygdala with other cortical areas, as seen in the medial and

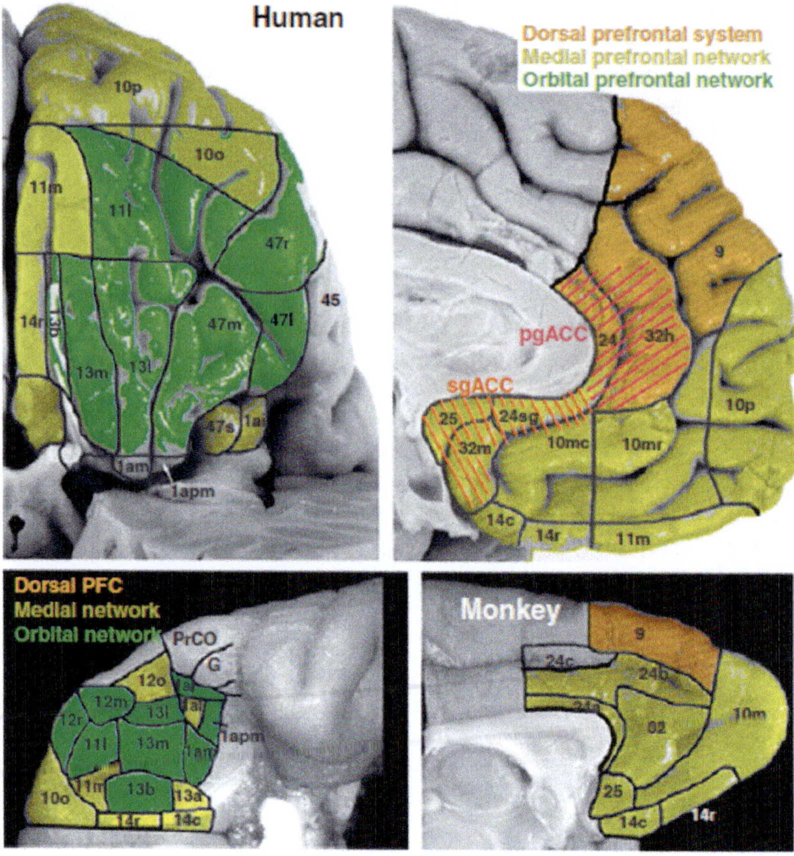

FIGURE 6.1 Maps of the human and monkey brain (frontal cortex). *Reproduced from the reference Price JL, Drevets WC. Neural circuits underlying the pathophysiology of mood disorders.* Trends Cogn Sci. *2012;16:61–71.*[21]

anterior temporal cortex and the posterior cingulate cortex. Anatomical evidence provided by animal studies (monkeys) suggest a connectional network, including those seen in several areas in the mPFC and the amygdala.[20,25] Also, connections were found with subcortical structure in the ventral striatum and pallidum, the hypothalamus, medial thalamus, and brainstem. These regions were termed as the limbic-cortical-striatal-pallidum-thalamus (LCSPT) circuit.

As seen in Fig. 6.3, it was found that degenerative basal ganglia diseases and lesions of striatum and orbitofrontal cortex increase the risk of developing MDD.[25] Due to neurological disorders, synaptic transmission along the LCSPT circuit was seen to be affected in multiple ways. Multiple types of dysfunctions in LCSPT circuits may cause depression.

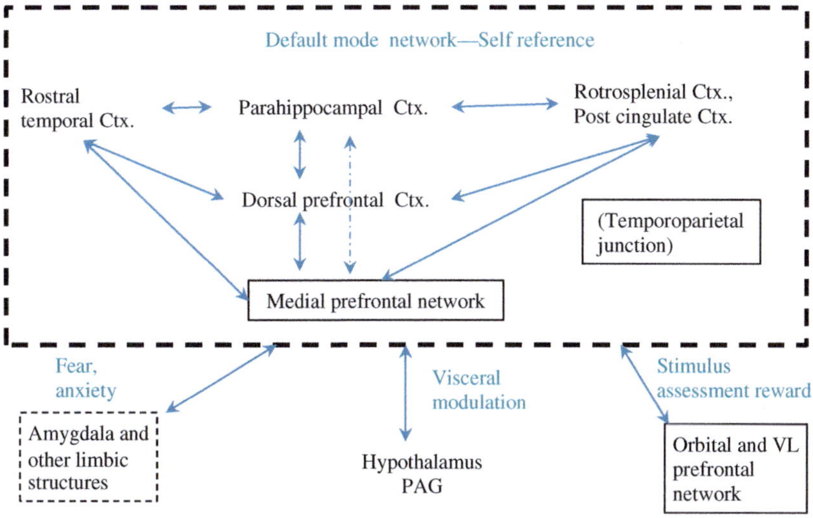

FIGURE 6.2 Brain areas implicated during MDD. *Reproduced from Price JL, Drevets WC. Neural circuits underlying the pathophysiology of mood disorders.* Trends Cogn Sci. *2012;16:61–71.*

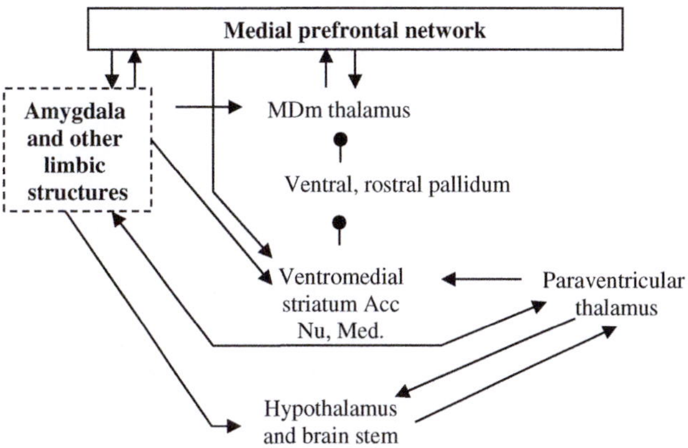

FIGURE 6.3 Brain areas identified from animal and human studies for MDD. *Reproduced from Price JL, Drevets WC. Neural circuits underlying the pathophysiology of mood disorders.* Trends Cogn Sci. *2012;16:61–71.*

6.2.3 Limbic Structures

The amygdala plays a central part in understanding emotion-related circuitry involving the limbic structure. Animal studies, including those performed on cats, rats, and monkeys, have elucidated that the medial prefrontal network has reciprocal connections to the amygdala, orbital network, insular and cortical areas, ventromedial striatum, and thalamic nucleus (please see Figs. 6.2 and 6.3 for illustration).[26]

Abnormalities associated with the glucose metabolism appear to be more specific to MDD subgroups. For example, the resting metabolism was found abnormally elevated in the left amygdala, specifically in depressed subjects diagnosed with bipolar depression (BD), familial pure depressive disease (FPDD), or the melancholic subtype.[9] In depressed subjects, amygdala showed exaggerated hemodynamic responses to stimuli such as explicit or implicit presentation of sad faces and backwardly masked fearful faces and blunted responses to masked happy faces.[22,23] Similarly, amygdala abnormalities were observed in unmedicated-remitted subjects with MDD.[22,24] However, antidepressant treatment shifts emotional processing biases toward positivity.[22,27,28] These evidences suggest that the amygdala and related circuitry are involved in the control of response biases during MDD.

6.2.4 Prefrontal Cortex

Axonal studies conducted during the 1990s thoroughly elucidated the subcortical and cortical circuits as related to the amygdala and orbitomedial prefrontal cortex (OMPFC). As shown in Fig. 6.1, the medial and dorsolateral prefrontal networks have similar connections with other cortical regions recognized within the OMPFC.[20] Similarly, in monkeys, three regions have been defined: the ventrolateral region ventral to the principal sulcus (VLPFC; related to the orbital prefrontal network), the dorsal prefrontal region dorsal to the principal sulcus (DPFC; like the medical prefrontal network), and a caudolateral region rostral to the arcuate sulcus (CLPFC).[29]

In an emotional go-no-go study, the hemodynamic activity of normal subjects was increased in the subgenual anterior cingulate cortex (sgACC) in response to stimuli such as sadness induction, exposure to traumatic reminders, and the selection of sad and happy targets.[10] On the contrary, remitted MDD subjects showed a decrease of coupling between the rostral superior temporal gyrus, hippocampus, hemodynamic responses of the sgACC, and the medial frontopolar cortex during guilt versus indignation.

The ventral pregenual anterior cingulate cortex (pgACC) and the ventromedial prefrontal cortex (vmPFC) have been implicated in MDD patients with anhedonia symptoms. In addition, the ventral pgACC showed increased activity in response to dopamine-inducing drugs during preference judgments.[30] On the other hand, depressed subjects exhibited reduced blood oxygen-level dependent (BOLD) activity during higher resting EEG delta current density and reward learning in association with anhedonia ratings.[30] In the supragenual ACC, attenuated BOLD responses were associated with a depressed state during the recall of autobiographical memories. In addition, these deficits were correlated to diminished activity in the parahippocampus and hippocampus.[31]

6.2.5 Cortical Projections to the Hypothalamus and Brainstem

According to animal studies, the medial prefrontal network connects with the hypothalamus, the periaqueductal gray, and different visceral control centers.[20] The subgenual cortex provides the heaviest projections terminated in the lateral and medial hypothalamus. The origins of the projections extend beyond the prefrontal network, including area nine in the DPFC and the rostral superior temporal gyrus.

According to observations in humans, activity in the mPFC correlates with visceral activation in response to emotional[32,33] and nonemotional stimuli.[34] In another study,[35] lesions in the mPFC were linked with severe or complete deficits in visceral responses to emotionally competent stimuli. These deficits have been linked to the absence of "somatic marker" provided by visceral activation. The overactivation of this visceromotor system may contribute to depression.

6.2.6 Cortico-Striatal-Thalamic Circuits Related to the Orbitomedial Prefrontal Cortex

As shown in Fig. 6.3, the PFC has specific connections with the thalamus and striatum. A well-known example is the cortico-striato-pallido-thalamic loops. The medial prefrontal network is connected to the medial segment of the ventromedial part of the striatum and the mediodorsal thalamic nucleus (MDm).[36,37]

In humans, neurophysiological activity involving the subcortical structures extensively connected with the medial prefrontal network was found to be correlated with depressive symptoms. For example, an elevation of metabolism in the accumbens area showed positive correlation with increasing anhedonia ratings.[38] In addition, hemodynamic responses of the ventral striatum to rewarding stimuli were found to decrease in depressed participants when compared with healthy controls. Moreover, increased levels of anhedonia were linked with blunted ventral striatal responses to rewarding stimuli in both healthy[30] and depressed subjects. Last, depressed participants exhibit impaired reward (but not punishment) reversal accuracy correlated with the attenuated ventral striatal BOLD response to unexpected reward.[39]

6.3 NEUROBIOLOGY OF ELECTROENCEPHALOGRAPHY-BASED PREDICTIVE BIOMARKER FOR DEPRESSION

This section elaborates on the meanings associated with EEG changes observed in the pretreatment and after medication. More specifically, neuronal mechanisms were observed that could elucidate or associated with such changes. More information on EEG-based predictions of treatment responses to antidepressants, termed as EEG-based biomarkers, can be found in Chapter 5 ("Using Electroencephalography for Diagnosing and Treating Depression").

6.3.1 Changes in Alpha Band Activity

Many studies have reported an increase in EEG alpha power because of treatment with tricyclic antidepressants (TCAs; including clomipramine and imipramine) and SSRIs (such as paroxetine and fluoxetine) which help differentiate between treatment respondents and non-respondents.[40–43] In this context, an increase in EEG alpha power has been termed as biomarker for ATR. In this section, what these changes in alpha frequency mean in the treatment of depression shall be addressed.

In the study,[44] an increase in EEG alpha power has been reported in depressed individuals who are treated with SSRIs. The authors assumed that elevated pretreatment alpha activity may be a marker of the resemblance between low arousal and low serotonergic activity. Low serotonergic activity could be caused by the diminished activity of the cortical afferents and mesencephalic raphe nuclei. A different study[45] has reported a biological mechanism that plays a key role in both increased alpha asymmetry and alpha power. The study has linked depression with an impairment of temporoparietal mechanisms that cause emotional arousal as well.

6.3.2 Electroencephalography Theta Band Activity

The review[46] has declared the existence of a robust connection between ATR and resting rACC activity. In addition, a neurobiological mechanism has been explained as well. The researchers declared rACC as a principal hub within the default network (DN) of the brain involved in self-focus processes.[47] In addition, the DN includes brain regions such as the dorsal medial PFC, ventromedial prefrontal cortex (PFC), retrosplenial cortex, posterior cingulate, lateral temporal cortex, lateral parietal cortex, and hippocampal formation. The study[48] associated elevated resting state activity with these DN regions, focusing on remembering, planning, and reflective thought, or task-independent introspection such as rumination. In particular, rumination is a mechanism of distress response.[49] Rumination is comprised of two components: brooding and reflective pondering.[50]

Reflective pondering is considered as an adaptive process that helps during cognitive tasks. On the contrary, brooding is analytic and involves self-focus. Ultimately, it is destructive and adds negative effects to depressive symptoms.[51] The study[46] has linked treatment response with elevated resting rACC activity through self-focus and adaptive self-referential functions (such as mindfulness). While handling external stimuli, not all brain regions are active. More specifically, the DN showed reduced activity. The resulting network is termed as the task negative network (TNN).[52] On the other hand, goal-oriented tasks involve cognitive control activities and attention that excites the task positive network (TPN), including the dorsal ACC, dorsolateral PFC, the middle temporal area, and the intraparietal sulcus.

During MDD, the relationship between the TPN and DN is impaired leading to brooding and maladaptive rumination and disrupted self-focus.[46] Because the DN has a functional connection with the limbic and paralimbic regions, impaired functional connection leads to impaired amygdala activity. Based on these facts, Pizzagalli[46] proposed that elevated rACC activity that corresponded with treatment response to antidepressants may play a major role in reconstructing the functional connections between the TRN and DN. In short, significant fronto-cingulate dysfunction has been reported in many areas of the study, including neuroimaging, neurophysiology, and neuropsychology.

6.3.3 Alpha Interhemispheric Asymmetry

Interhemispheric asymmetry is considered as a marker for depression. More specifically, alpha band asymmetry characterized by left lateralization has been reported during treatment response involving TCAs and SSRIs.[40,44,53] These studies have commonly reported alpha asymmetry as a predictor of treatment outcome at 4–12 weeks post-treatment. There is an opposite relationship between cortical activity and alpha power; for example, reduced alpha implicates increased activity and increased alpha means reduced cortical activity.

Treatment response has been linked with substantial left rather than right hemispheric activation. On the contrary, nonresponders have shown opposite patterns, that is, pronounced right rather than left activations in posterior and frontal regions.[44] In the study,[54] reduced left frontal activity was reported that is in conflict to the finding of left lateralization. In a study,[55] Bruder et al. has explained this conflict by recommending that lessened left prefrontal activation may disinhibit left temporoparietal regions manifesting as increased left hemisphere advantage in treatment (fluoxetine) responder.

Studies[56,57] have supported this hypothesis of inhibitory association between temporoparietal and frontal regions. In a study,[58] Devidson and Henrique suggest that left frontal hypoactivation can be explained as a shortfall in approach mechanisms and right frontal hypoactivation may indicate deficits in withdrawal mechanisms.

In treatment responders, alpha asymmetry can be trait-dependent because it is found in euthymic patients as well as in depressed adolescents and adults.[59–61] In addition, asymmetry has been found in both adult and young descendants of depressed parents.[62,63] In conclusion, the alpha interhemispheric asymmetry could implicate the underlying pathophysiology as a trait marker for depression. In addition, it could be implicated as the vulnerability to inherite depression that responded to serotonergic agents.[44]

There are few attempts to explain alpha asymmetry in the light of treatment response. For example, Bruder theorized that the distribution of the 5-HT neurotransmitter system may have a lateralized nature inside the brain and could be asymmetrically distributed in depressed patients or a subtype of depressed patients.[40] The 5-HT asymmetric distribution is an important finding because it is not only implicated in mood disruption, but is also affected by medication, including antidepressants.

A few studies have replicated the theory that 5-HT pathways are asymmetric for opposite sides of the brain,[64,65] however, others could not replicate this finding.[65,66] Another hypothesis proposed by Bruder stated that alpha asymmetry in SSRI responders may be due to low arousal associated with low serotonergic activity.[44] Based on these evidences, Alhaj et al. have suggested that gene polymorphisms of the 5-HT1A receptor may have association with alpha asymmetry.[67] Hence, 5-HT1A receptor polymorphisms largely corresponded with the pathophysiology of depression. Moreover, the 5-HT transporter is one of the primary targets of antidepressant medication.[68,69]

6.3.4 Theta Cordance

Theta cordance provides a mathematical formula that combines information from both relative and absolute power involving the EEG theta spectra.[70] Theta cordance (a combination of both absolute and relative power) has shown more association with regional cerebral perfusion than absolute and relative power alone.[71] In addition, cordance is robust against factors such as age, gender, and severity of baseline depression.[72] Studies[73–75] have reported that the decreased frontal cordance measured from depressed patients over 48 hours to 1-week posttreatment have shown prediction of treatment response with an accuracy of 72%–88%. While the study participants were treated with both serotonin-norepinephrine reuptake inhibitors (SNRI) (venlafaxine) and SSRIs (fluoxetine).

Although theta cordance has shown association with treatment outcome, its associated brain mechanism remains unclear. Most EEG-derived biomarkers for depression previously mentioned have demonstrated dissimilarities in brain activity at baseline prior to treatment; the cordance does not. Significant differences can be observed for 2–7 days posttreatment. This was explained as early changes in brain activity because of antidepressant medication.[76]

Unfortunately, no clear brain mechanism has been proposed for theta cordance. One possible explanation could be the fact that theta cordance involves theta band activity; therefore the explanation of theta band provided by Pizzagalli[46] could be a possibility in this context as well.

6.3.5 Antidepressant Treatment Response Index

Like QEEG theta cordance, the ATR index is a mathematical combination of frontal EEG alpha and theta power. Observations of frontal alpha and theta powers are performed at pretreatment and at 1-week after treatment starts. The Biomarkers for Rapid Identification of Treatment Effectiveness in Major Depression (BRITE-MD) study[77] examined the utility of the ATR index as a biological marker of treatment response for depression treatment. In this study, depressed patients were treated for 1 week with escitalopram (ESC) and then randomized to continue ESC or switch to either bupropion (BUP) or its augmentation with ESC. The study concluded with an accuracy of 74% in predicting response and remission. The study also explored the capability of the ATR index as a differential response predictor to either ESC or BUP monotherapy.

The ATR index works by setting an empirically derived threshold; for example, the ROC curve for predictive accuracy was drawn and an optimal threshold of 58.6 was chosen. The importance of this threshold was demonstrated by the fact that those patients having high ATR values (above the threshold value) showed 2.4 times higher likelihood to respond to ESC than those having lower ATR values (below the threshold value). The accuracies were 68% and 28%, respectively ($P = .001$).

In addition, patients with ATR values lower than the threshold were converted to BUP treatment and were 1.9 times more responsive to BUP alone when compared with those patients who remained on ESC treatment. The prediction accuracies were 53% and 28%, respectively

($P = .034$). Hence, it was concluded that the ATR index may be a useful tool in predicting the treatment outcome and could be used in guiding the treatment decision. However, from a machine learning point of view, accuracies such as 53% are considered low and that could explain why the ATR index cannot be utilized further for clinical applications. Like QEEG theta cordance, the ATR index has showed promise as a marker for treatment outcome prediction for depression; however there remains a gap in explaining the biological mechanisms associated with it.

Due to this reason, researchers have questioned the practical implications of the ATR index.[78] In response to such concerns, Leuchter[79] explained a possible mechanism that could be associated with it. According to Leuchter and his team and in accordance with the BRITE-MD study, it has been theorized that both the QEEG theta cordance and the ATR index reflect activity arising from frontal rhythmic activity. This activity is assumed to originate from the PFC and ACC, which are also involved in the pathophysiology of depression.[80]

In short, the ATR index and QEEG theta cordance were associated with early functional changes in both the ACC and PFC. The ATR index is based on a combination of both the theta and alpha activities. Therefore, the neurobiology explained by theta and alpha frequencies could also be refered to explain the neurobiology of the ATR index. Unfortunately, the differential responses reported during ATR-based studies largely remain unexplored and unexplained. This warrants further investigation.

6.3.6 The P300

The P300 is an event-related potential (ERP) that is computed based on grand averaging various epochs related to either targets or nontargets in visual or audio stimulus tasks. Because of the grand averaging, a positive peak termed as P300 computed and appeared at 300 ms after the experimental visual stimulus. Researchers have linked attention and auditory processes with the P300 index.[81,82] During depression, a delayed P300 component (also known as latency) has been observed.[83] On the contrary, delayed latency was reported to normalize with antidepressant treatment after 4 weeks.[84] The relationship between P300 and antidepressant treatment was further studied with these conclusions. For example, the higher P300 amplitudes at occipital sites were associated with treatment response while the patients were treated with fluoxetine and TCA.[85] In addition, a study involving elderly MDD patients reported association of longer P300 latency with patients who did not remit.[86]

The exact neurobiology associated with P300 latency and amplitude is unknown except for a few explanations such as P300 latency being correlated with prolactin response to the 5-HT1A agonist flesinoxan.[87] This explanation seems interesting because the 5-HT1A receptor is vital during the therapeutic action of antidepressants.

6.3.7 LDEAP

P300 shows a positive peak before 300 ms after the occurrence of auditory stimuli. In addition, N1 and P2 manifest negative peaks at 100 ms and a positive peak at 200 ms accordingly. LDEAP involves N1/P2 components observed in response to auditory stimuli. The LDEAP amplitude changes due to increases in the volume of the stimulus. According to a study, the neurobiology of LDEAP has been associated with the 5-HT neurotransmission system, observed in the auditory cortex.[88]

In the study,[89] a larger slope of P2 amplitude as a function of stimulus intensity at baseline was correlated with treatment response to SSRIs. Similar findings were replicated for bupropion as well.[90] Empirical findings based on LDAEP suggest dividing the study samples into the top 50% range (higher slopes; "strong" LDAEP) and the bottom 50% range (lower slopes; "weak" LDAEP) based on median split midpoint. These studies determined an association of strong LDAEP at baseline with response to SSRIs such as citalopram, paroxetine, and fluoxetine.[91–93] On the contrary, weak LDAEP at baseline was correlated with response to reboxetine; the norepinephrine reuptake inhibitor (NRI).[92,94,95]

In short, LDAEP may be a differential marker for ATR with distinct mechanisms of action (SSRIs and NRIs). As LDAEP has a hypothesized link with 5-HT activity, it could be considered as a differential biomarker for depression. In their study,[96] the amplitude of the slope of a plot of N1/P2 against loudness (the amplitude/stimulus intensity function [ASF-slope]) was inversely associated with serotonergic activity. In other words, the suppression of ERP responsiveness to auditory tone intensity was because of the high level of 5-HT, whereas the facilitation of ERP responsiveness was due to low 5-HT. In addition, the ASF-slope has been negatively correlated with plasma 5-HT concentration following administration of SSRI (fluvoxamine).[88] This finding was reported in depressed patients only.

A review by O'Neill[97] concluded that the LDAEP may not be a good index for 5-HT. Overall, the review supported the idea of using LDAEP as a biomarker for treatment response for depression. Table 6.1 summarizes the key points of EEG-based biomarkers and their predictive relationship with treatment response of antidepressants.

TABLE 6.1 EEG-Derived Biomarkers and Their Correlation with Antidepressant Treatment Response

EEG Biomarkers	Response Type	Time of Treatment	Antidepressant Type	Biological Implications
Activation of the alpha band	Elevated	Pretreatment	SSRI and TCA	Right temporoparietal region; low 5-HT
Activation of the theta band	Elevated	Pretreatment	NRI, SSRI, SNRI, and TCA	rACC and DN
Asymmetry of the alpha band	Left Lateralization	Pretreatment	TCA and SSRI	HT1A polymorphismLateralized 5-HT system; low 5-HT
Theta cordance	Decreased	2–7 days posttreatment	SNRI and SSRI	ACC and PFC
The ATR index	High	Pretreatment and 1 week PT	SSRI	ACC and PFC
	Low	Pretreatment and 1 week PT	NDRI	
P300	High amplitude	Pretreatment	SSRI and TCA	5-HT1A receptor
LDEAP	High amplitude	Pretreatment	SSRI and NDRI	5-HT activity in the CNS
	Strong LDAEP	Pretreatment	SSRI	
	Weak LDAEP	Pretreatment	NRI	

TABLE 6.2 Anatomical Abnormalities During Depression

Brain Dysfunctions	Effected Brain Regions	Critical Analysis
Degenerative diseases and lesions (effected synaptic transmission in the LCSPT)[30,31] (modality = SPECT)	Striatum, orbitofrontal cortex (OFC)	MDD has been reported as a comorbid disease and brain regions identified during depression are not unique identifiers specific to depression. Hence, the discrimination and subjectivity associated with pathophysiology make it difficult to find exact causes of depression.
Abnormal glucose metabolism[38] (modality = fMRI)	Amygdala	
Decreased hemodynamic responses[39] (modality = fMRI)	sgACC, rostral superior temporal gyrus, hippocampus, and medial frontopolar cortex	
Reduced BOLD responses[39] (modality = fMRI)	Ventral pgACC, supragenual ACC	
Anhedonia[31] (modality = fMRI)	PgACC, VmPFC	

6.4 SUMMARY

Correlations between depressive symptoms and neurophysiological activities have been found to be common in subcortical structures and the medial prefrontal network. In the accumbens area, a rise in metabolism under catecholamine depletion was positively correlated with anhedonia ratings.[38] In addition, depressed subjects showed decreased hemodynamic responses to reward stimuli in the ventral striatum. High levels of anhedonia were correlated with blunted ventral striatal responses to reward stimuli in both depressed and healthy subjects.[30] Depressed subjects showed impaired reward processing during probabilistic reversal learning. Moreover, the learning was found correlated with attenuated ventral striatal BOLD response to unexpected reward stimuli.[39] Table 6.2 provides a summary of anatomical abnormalities found during depression.

References

1. American Psychiatric Association. Diagnostic and statistical manual of mental disorders (DSM-5®). *American Psychiatric Pub*, 2013.
2. Treadway MT, Zald DH. Reconsidering anhedonia in depression: lessons from translational neuroscience. *Neurosci Biobehav Rev*. 2011;35:537–555.
3. Henriques JB, Glowacki JM, Davidson RJ. Reward fails to alter response bias in depression. *J Abnorm Psychol*. 1994;103:460.

4. Pizzagalli DA, Iosifescu D, Hallett LA, Ratner KG, Fava M. Reduced hedonic capacity in major depressive disorder: evidence from a probabilistic reward task. *J Psychiatr Res*. 2008;43:76–87.
5. Pizzagalli DA, Jahn AL, O'Shea JP. Toward an objective characterization of an anhedonic phenotype: a signal-detection approach. *Biol Psychiatry*. 2005;57:319–327.
6. Tavares JT, Drevets W, Sahakian B. Cognition in mania and depression. *Psychol Med*. 2003;33:959–967.
7. Roiser JP, Cannon DM, Gandhi SK, et al. Hot and cold cognition in unmedicated depressed subjects with bipolar disorder. *Bipolar Disord*. 2009;11:178–189.
8. Tavares JVT, Clark L, Furey ML, Williams GB, Sahakian BJ, Drevets WC. Neural basis of abnormal response to negative feedback in unmedicated mood disorders. *Neuroimage*. 2008;42:1118–1126.
9. Drevets WC, Price JL, Furey ML. Brain structural and functional abnormalities in mood disorders: implications for neurocircuitry models of depression. *Brain Struct Funct*. 2008;213:93–118.
10. Elliott R, Rubinsztein JS, Sahakian BJ, Dolan RJ. Selective attention to emotional stimuli in a verbal go/no-go task: an fMRI study. *Neuroreport*. 2000;11:1739–1744.
11. Murphy F, Sahakian B, Rubinsztein J, et al. Emotional bias and inhibitory control processes in mania and depression. *Psychol Med*. 1999;29:1307–1321.
12. Murray LA. Mood congruence and depressive deficits in memory: a forced-recall analysis. *Memory*. 1999;7:175–196.
13. Bradley BP, Mogg K, Williams R. Implicit and explicit memory for emotion-congruent information in clinical depression and anxiety. *Behav Res Ther*. 1995;33:755–770.
14. Broomfield NM, Davies R, MacMahon K, Ali F, Cross SM. Further evidence of attention bias for negative information in late life depression. *Int J Geriatr Psychiatry*. 2007;22:175–180.
15. Pérez MG, Rivera RMB, Fuster AB, Rodríguez MÁR. Attentional biases and vulnerability to depression. *Span J Psychol*. 1999;2:11–19.
16. Erickson K, Drevets WC, Clark L, et al. Mood-congruent bias in affective go/no-go performance of unmedicated patients with major depressive disorder. *Am J Psychiatry*. 2005;162:2171–2173.
17. Gotlib IH, Kasch KL, Traill S, Joormann J, Arnow BA, Johnson SL. Coherence and specificity of information-processing biases in depression and social phobia. *J Abnorm Psychol*. 2004;113:386.
18. Gotlib IH, Krasnoperova E, Yue DN, Joormann J. Attentional biases for negative interpersonal stimuli in clinical depression. *J Abnorm Psychol*. 2004;113:127.
19. Price JL, Drevets WC. Neurocircuitry of mood disorders. *Neuropsychopharmacology*. 2010;35:192.
20. Öngür D, Price J. The organization of networks within the orbital and medial prefrontal cortex of rats, monkeys and humans. *Cereb Cortex*. 2000;10:206–219.
21. Price JL, Drevets WC. Neural circuits underlying the pathophysiology of mood disorders. *Trends Cogn Sci*. 2012;16:61–71.
22. Victor TA, Furey ML, Fromm SJ, Öhman A, Drevets WC. Relationship of emotional processing to masked faces in the amygdala to mood state and treatment in major depressive disorder. *Arch Gen Psychiat*. 2010;67:1128.
23. Suslow T, Konrad C, Kugel H, et al. Automatic mood-congruent amygdala responses to masked facial expressions in major depression. *Biol Psychiatry*. 2010;67:155–160.
24. Neumeister A, Charney DS, Belfer I, et al. Sympathoneural and adrenomedullary functional effects of [alpha] 2C-adrenoreceptor gene polymorphism in healthy humans. *Pharmacogenet Genom*. 2005;15:143–149.

25. Price JL, Drevets WC. Neurocircuitry of mood disorders. *Neuropsychopharmacology*. 2009;35:192–216.
26. Price JL. Comparative aspects of amygdala connectivity. *Ann N Y Acad Sci*. 2003;985:50–58.
27. Harmer CJ. Serotonin and emotional processing: does it help explain antidepressant drug action? *Neuropharmacology*. 2008;55:1023–1028.
28. Killgore WD, Yurgelun-Todd DA. Activation of the amygdala and anterior cingulate during nonconscious processing of sad versus happy faces. *Neuroimage*. 2004;21:1215–1223.
29. Saleem KS, Kondo H, Price JL. Complementary circuits connecting the orbital and medial prefrontal networks with the temporal, insular, and opercular cortex in the macaque monkey. *J Comp Neurol*. 2008;506:659–693.
30. Wacker J, Dillon DG, Pizzagalli DA. The role of the nucleus accumbens and rostral anterior cingulate cortex in anhedonia: integration of resting EEG, fMRI, and volumetric techniques. *NeuroImage*. 2009;46:327–337.
31. Young K, Erickson K, Nugent A, et al. Functional anatomy of autobiographical memory recall deficits in depression. *Psychol Med*. 2012;42:345–357.
32. Critchley HD, Elliott R, Mathias CJ, Dolan RJ. Neural activity relating to generation and representation of galvanic skin conductance responses: a functional magnetic resonance imaging study. *J Neurosci*. 2000;20:3033–3040.
33. Williams LM, Brammer MJ, Skerrett D, et al. The neural correlates of orienting: an integration of fMRI and skin conductance orienting. *Neuroreport*. 2000;11:3011–3015.
34. Teves D, Videen TO, Cryer PE, Powers WJ. Activation of human medial prefrontal cortex during autonomic responses to hypoglycemia. *Proc Natl Acad Sci*. 2004;101:6217–6221.
35. Bechara A, Damasio H, Tranel D, Damasio AR. The Iowa Gambling Task and the somatic marker hypothesis: some questions and answers. *Trends Cogn Sci*. 2005;9:159–162.
36. Ray JP, Price JL. The organization of projections from the mediodorsal nucleus of the thalamus to orbital and medial prefrontal cortex in macaque monkeys. *J Comp Neurol*. 1993;337:1–31.
37. Ferry AT, Öngür D, An X, Price JL. Prefrontal cortical projections to the striatum in macaque monkeys: evidence for an organization related to prefrontal networks. *J Comp Neurol*. 2000;425:447–470.
38. Hasler G, Fromm S, Carlson PJ, et al. Neural response to catecholamine depletion in unmedicated subjects with major depressive disorder in remission and healthy subjects. *Arch Gen Psychiat*. 2008;65:521–531.
39. Robinson OJ, Cools R, Carlisi CO, Sahakian BJ, Drevets WC. Ventral striatum response during reward and punishment reversal learning in unmedicated major depressive disorder. *Am J Psychiatry*. 2012;169:152–159.
40. Bruder GE, Stewart JW, Tenke CE, et al. Electroencephalographic and perceptual asymmetry differences between responders and nonresponders to an SSRI antidepressant22. *Biol Psychiatry*. 2001;49:416–425.
41. Knott VJ, Telner JI, Lapierre YD, Browne M, Horn ER. Quantitative EEG in the prediction of antidepressant response to imipramine. *J Affect Disord*. 1996;39:175–184.
42. Knott V, Mahoney C, Kennedy S, Evans K. Pre-treatment EEG and it's relationship to depression severity and paroxetine treatment outcome. *Pharmacopsychiatry*. 2000;33:201–205.
43. Ulrich G, Haug H-J, Stieglitz R-D, Fähndrich E. EEG characteristics of clinically defined on-drug-responders and non-responders-a comparison clomipramine vs. maprotiline. *Pharmacopsychiatry*. 1988;21:367–368.

44. Bruder GE, Sedoruk JP, Stewart JW, McGrath PJ, Quitkin FM, Tenke CE. Electroencephalographic alpha measures predict therapeutic response to a selective serotonin reuptake inhibitor antidepressant: pre-and post-treatment findings. *Biol Psychiatry*. 2008;63:1171–1177.
45. Heller W, Etienne MA, Miller GA. Patterns of perceptual asymmetry in depression and anxiety: implications for neuropsychological models of emotion and psychopathology. *J Abnorm Psychol*. 1995;104:327.
46. Pizzagalli DA. Frontocingulate dysfunction in depression: toward biomarkers of treatment response. *Neuropsychopharmacology*. 2011;36:183.
47. Buckner R, Andrews-Hanna J, Schacter D. The brain's default network: anatomy, function, and relevance to disease,. *Ann N Y Acad Sci*. 2008;1124:1–38.
48. Simpson JR, Snyder AZ, Gusnard DA, Raichle ME. Emotion-induced changes in human medial prefrontal cortex: I. During cognitive task performance. *Proc Natl Acad Sci*. 2001;98:683–687.
49. Nolen-Hoeksema S, Wisco BE, Lyubomirsky S. Rethinking rumination. *Perspect Psychol Sci*. 2008;3:400–424.
50. Treynor W, Gonzalez R, Nolen-Hoeksema S. Rumination reconsidered: a psychometric analysis. *Cognit Ther Res*. 2003;27:247–259.
51. Watkins E, Teasdale JD. Adaptive and maladaptive self-focus in depression. *J Affect Disord*. 2004;82:1–8.
52. Shulman GL, Fiez JA, Corbetta M, et al. Common blood flow changes across visual tasks: II. Decreases in cerebral cortex. *J Cogn Neurosci*. 1997;9:648–663.
53. Ulrich G, Renfordt E, Zeller G, Frick K. Interrelation between changes in the EEG and psychopathology under pharmacotherapy for endogenous depression. *Pharmacopsychiatry*. 1984;17:178–183.
54. Henriques JB, Davidson RJ. Left frontal hypoactivation in depression. *J Abnorm Psychol*. 1991;100:535.
55. Bruder GE, Otto MW, McGrath PJ, et al. Dichotic listening before and after fluoxetine treatment for major depression: relations of laterality to therapeutic response. *Neuropsychopharmacology*. 1996;15:171.
56. Knight RT, Hillyard SA, Woods DL, Neville HJ. The effects of frontal and temporal-parietal lesions on the auditory evoked potential in man. *Electroencephalogr Clin Neurophysiol*. 1980;50:112–124.
57. Tucker M, Stenslie CE, Roth RS, Shearer SL. Right frontal lobe activation and right hemisphere performance: decrement during a depressed mood. *Arch Gen Psychiat*. 1981;38:169–174.
58. Davidson RJ, Henriques J. Regional brain function in sadness and depression. *Neuropsychol Emot*. 2000;269–297.
59. Henriques JB, Davidson RJ. Regional brain electrical asymmetries discriminate between previously depressed and healthy control subjects. *J Abnorm Psychol*. 1990;99:22.
60. Kentgen LM, Tenke CE, Pine DS, Fong R, Klein RG, Bruder GE. Electroencephalographic asymmetries in adolescents with major depression: influence of comorbidity with anxiety disorders. *J Abnorm Psychol*. 2000;109:797.
61. Reid SA, Duke LM, Allen JJ. Resting frontal electroencephalographic asymmetry in depression: inconsistencies suggest the need to identify mediating factors. *Psychophysiology*. 1998;35:389–404.
62. Bruder GE, Tenke CE, Warner V, et al. Electroencephalographic measures of regional hemispheric activity in offspring at risk for depressive disorders. *Biol Psychiatry*. 2005;57:328–335.
63. Jones NA, Field T, Fox NA, Lundy B, Davalos M. EEG activation in 1-month-old infants of depressed mothers. *Dev Psychopathol*. 1997;9:491–505.

64. Mandell AJ, Knapp S. Asymmetry and mood, emergent properties of serotonin regulation: a proposed mechanism of action of lithium. *Arch Gen Psychiat.* 1979;36:909–916.
65. Tucker DM, Williamson PA. Asymmetric neural control systems in human self-regulation. *Psychol Rev.* 1984;91:185.
66. Arato M, Frecska E, MacCrimmon DJ, et al. Serotonergic interhemispheric asymmetry: neurochemical and pharmaco-EEG evidence. *Prog Neuropsychopharmacol Biol Psychiatry.* 1991;15:759–764.
67. Bismark AW, Moreno FA, Stewart JL, et al. Polymorphisms of the HTR1a allele are linked to frontal brain electrical asymmetry. *Biol Psychol.* 2010;83:153–158.
68. Blier P, De Montigny C. Current advances and trends in the treatment of depression. *Trends Pharmacol Sci.* 1994;15:220–226.
69. Savitz J, Lucki I, Drevets WC. 5-HT1A receptor function in major depressive disorder. *Prog Neurobiol.* 2009;88:17–31.
70. Leuchter AF, Cook IA, Lufkin RB, et al. Cordance: a new method for assessment of cerebral perfusion and metabolism using quantitative electroencephalography. *Neuroimage.* 1994;1:208–219.
71. Leuchter AF, Uijtdehaage SH, Cook IA, O'Hara R, Mandelkern M. Relationship between brain electrical activity and cortical perfusion in normal subjects. *Psychiatry Res. Neuroimaging.* 1999;90:125–140.
72. Morgan ML, Witte EA, Cook IA, Leuchter AF, Abrams M, Siegman B. Influence of age, gender, health status, and depression on quantitative EEG. *Neuropsychobiology.* 2005;52:71–76.
73. Bares M, Brunovsky M, Kopecek M, et al. Early reduction in prefrontal theta QEEG cordance value predicts response to venlafaxine treatment in patients with resistant depressive disorder. *Eur Psychiatry.* 2008;23:350–355.
74. Cook IA, Leuchter AF, Morgan M, et al. Early changes in prefrontal activity characterize clinical responders to antidepressants. *Neuropsychopharmacology.* 2002;27:120–131.
75. Cook IA, Leuchter A. Prefrontal changes and treatment response prediction in depression. *Semin Clin Neuropsychiatry.* 2001;113–120.
76. Iosifescu DV. Electroencephalography-derived biomarkers of antidepressant response. *Harv Rev Psychiatry.* 2011;19:144–154.
77. Leuchter AF, Cook IA, Marangell LB, et al. Comparative effectiveness of biomarkers and clinical indicators for predicting outcomes of SSRI treatment in major depressive disorder: results of the BRITE-MD study. *Psychiatry Res.* 2009;169:124–131.
78. Kuo C-C, Tsai J-F. Cordance or antidepressant treatment response (ATR) index? *Psychiatry Res.* 2010;180:60.
79. Leuchter AF, Cook IA. Reply to Kuo and Tsai. *Psychiatry Res.* 2010;180:61–62.
80. Asada H, Fukuda Y, Tsunoda S, Yamaguchi M, Tonoike M. Frontal midline theta rhythms reflect alternative activation of prefrontal cortex and anterior cingulate cortex in humans. *Neurosci Lett.* 1999;274:29–32.
81. Mulert C, Pogarell O, Juckel G, et al. The neural basis of the P300 potential. *Eur Arch Psychiatry Clin Neurosci.* 2004;254:190–198.
82. Volpe U, Mucci A, Bucci P, Merlotti E, Galderisi S, Maj M. The cortical generators of P3a and P3b: a LORETA study. *Brain Res Bull.* 2007;73:220–230.
83. Bruder GE, Towey JP, Stewart JW, Friedman D, Tenke C, Quitkin FM. Event-related potentials in depression: influence of task, stimulus hemifield and clinical features on P3 latency. *Biol Psychiatry.* 1991;30:233–246.
84. Hetzel G, Moeller O, Evers S, et al. The astroglial protein S100B and visually evoked event-related potentials before and after antidepressant treatment. *Psychopharmacology.* 2005;178:161–166.

85. Bruder GE, Tenke CE, Stewart JW, et al. Brain event-related potentials to complex tones in depressed patients: relations to perceptual asymmetry and clinical features. *Psychophysiology*. 1995;32:373–381.
86. Kalayam B, Alexopoulos GS. Prefrontal dysfunction and treatment response in geriatric depression. *Arch Gen Psychiat*. 1999;56:713–718.
87. Hansenne M, Ansseau M. P300 event-related potential and serotonin-1A activity in depression. *Eur Psychiatry*. 1999;14:143–147.
88. Hegerl U, Gallinat J, Juckel G. Event-related potentials: do they reflect central serotonergic neurotransmission and do they predict clinical response to serotonin agonists? *J Affect Disord*. 2001;62:93–100.
89. Paige SR, Fitzpatrick DF, Kline JP, Balogh SE, Hendricks SE. Event-related potential amplitude/intensity slopes predict response to antidepressants. *Neuropsychobiology*. 1994;30:197–201.
90. Paige SR, Hendricks SE, Fitzpatrick DF, Balogh S, Burke WJ. Amplitude/intensity functions of auditory event-related potentials predict responsiveness to bupropion in major depressive disorder. *Psychopharmacol Bull*. 1995.
91. Gallinat J, Bottlender R, Juckel G, et al. The loudness dependency of the auditory evoked N1/P2-component as a predictor of the acute SSRI response in depression. *Psychopharmacology*. 2000;148:404–411.
92. Juckel G, Pogarell O, Augustin H, et al. Differential prediction of first clinical response to serotonergic and noradrenergic antidepressants using the loudness dependence of auditory evoked potentials in patients with major depressive disorder. *J Clin Psychiatry*. 2007;68:1206–1212.
93. Lee T-W, Yu YW, Chen T-J, Tsai S-J. Loudness dependence of the auditory evoked potential and response to antidepressants in Chinese patients with major depression. *J Psychiatry Neurosci*. 2005;30:202.
94. Linka T, Müller B, Bender S, Sartory G, Gastpar M. The intensity dependence of auditory evoked ERP components predicts responsiveness to reboxetine treatment in major depression. *Pharmacopsychiatry*. 2005;38:139–143.
95. Mulert C, Juckel G, Brunnmeier M, et al. Prediction of treatment response in major depression: integration of concepts. *J Affect Disord*. 2007;98:215–225.
96. Hegerl U, Juckel G. Intensity dependence of auditory evoked potentials as an indicator of central serotonergic neurotransmission: a new hypothesis. *Biol Psychiatry*. 1993;33:173–187.
97. O'neill BV, Croft RJ, Nathan PJ. The loudness dependence of the auditory evoked potential (LDAEP) as an in vivo biomarker of central serotonergic function in humans: rationale, evaluation and review of findings. *Hum Psychopharmacol Clin Exp*. 2008;23:355–370.

CHAPTER

7

Design of an Electroencephalography Experiment for Assessing Major Depressive Disorder

7.1 INTRODUCTION

This chapter elaborates on the experimental design involving electroencephalography (EEG) data acquisition. A properly designed experiment is fundamental and holds a central position in clinical studies. On the contrary, a poorly designed experiment may result in low quality results and false inferences. This could be more critical in the case of clinical studies than in normal scenarios. In this chapter, the experimental design of an EEG-based study is explained by designing an experiment for EEG-based diagnosis and prediction of treatment outcomes involving depression.

In general, an EEG-based experiment involves the design of study protocol, sample size calculation, recruitment criteria for study participants, the selection of appropriate questionnaires utilized for clinical assessment (e.g., depression), and EEG/ERP experimental procedures. EEG/ERP experiment design has been explained in detail in this chapter, including the experimental setup, the hardware system utilized, and the recording software.

In addition, the chapter elaborates on methods such as the construction of topographic maps. Maps were constructed to visualize abnormalities caused by depression. The localization of abnormalities was possible based on the significant differences between study groups. Moreover, to investigate the interrelation between clinical measures and EEG features, the Pearson correlation was performed.

EEG-Based Experiment Design for Major Depressive Disorder
DOI: https://doi.org/10.1016/B978-0-12-817420-3.00007-2

The rest of this chapter explains the design of study protocol, including sample size calculation, recruitment criteria, selected clinical questionnaires for quantitative analysis, and experimental setup for EEG data acquisition as well as the study's participant information gathering which could be helpful during EEG analysis. Finally, the chapter explains EEG localization techniques such as topographic plots, EEG/ERP differences between major depressive disorder (MDD) patients and healthy controls.

7.2 DESIGN OF STUDY PROTOCOL

This section describes the design of study protocol, including sample size calculation, recruitment criteria for study participants, questionnaires for clinical data acquisition, and EEG/ERP experiment design. In this study, experimental data collection involved human subjects as participants. Therefore, to ensure safety and confidentiality, standard scientific methods were adopted such as the design of study protocol, including sample size calculation, design of recruitment criteria, and experiment for EEG/ERP data acquisition. The approval of ethics for this study is attached in Appendix D. A brief description of each step is provided in Sections 7.2.1–7.2.4.

7.2.1 Sample Size Calculation

In this study, a group of 34 MDD patients were recruited based on the mathematical formula provided in Eq. (7.1)[1,2]:

$$n = \frac{P(1-P) \cdot (Z_{1-\alpha/2})^2}{e^2} \tag{7.1}$$

where P is the expected proportion (e.g., the expected diagnostic sensitivity), e is the error limit which is one half of the desired width of the confidence interval, and $Z_{1\text{-}\alpha/2}$ is the standard normal z-value corresponding to a cumulative probability of $1-\alpha/2$. The investigator must specify the best estimate for the proportion that is expected to be found after performing the study.[2]

Table 7.1 shows parameter values used to calculate the sample size for this study. The significance value (α) was assigned a value of 5% because it should be small as it indicates the probability of making an error. In addition, it implies a 5% chance of making an error due to misclassification. Furthermore, the power of the test (β) should be 80%, as this was considered fair enough to prove confidence in the decision of a correct classification. In this study, the value of P was assigned as 90%

TABLE 7.1 Parameters for Sample Size Calculation

Parameters and Assigned Values	Sample Size Calculation
Significance $\alpha = .05$ (alpha)	$P = .90, \alpha = 0.05, e = 0.10, Z_{1-\alpha/2} = 1.96$
Power of test = 80%, $\beta = .2$ (beta)	$n = \frac{(0.90)(0.10)(1.96)^2}{(0.10)^2} \cong 34$
Expected diagnostic accuracy $P = 90\%$,[4] expected error $e = 10\%$	

because a relevant study published previously had reported a classification accuracy of 87.5%.[3] Finally, the expected error was set to 10%, which showed that the method trained with the computed sample size may show a deviation of accuracy. After sample size calculations, the sample size is 34 MDD and 34 healthy controls.

7.2.2 Recruitment Criteria

Participation for this study was voluntary and each participant had agreed to participate willingly by signing a consent form of participation. In particular, MDD patients with a confirmed diagnosis were included and those MDD patients who could sign the consent form of participation. The MDD patients and healthy controls were well-informed regarding the study protocol and could leave the study any time during its progression without any written notification. The safety and security of the information regarding each study participant was ensured and intact.

In this study, participants from all ages were included because the MDD prevails in all ages starting from 18 to 65. The reason for the exclusion of people of other ages was that infants and older people would have been unable to perform the EEG experiment, which required frequent visits to the clinic. Only those MDD patients who had not yet started their medication were included. Based on expert opinion, MDD patients who switched their medication from one class of antidepressants to another class were also able to participate. In such a case, it was ensured that the MDD patients recruited for this study would have a washout time of at least 2 weeks. In addition, MDD patients with non-psychotic symptoms were excluded as to keep a low variability among the study participants. To further decrease variability due to external factors, patients with drug abuse, pregnant patients, and those with epilepsy were excluded. Table 7.2 summarizes exclusion and inclusion criteria for this research work.

TABLE 7.2 Study Inclusion and Exclusion Criteria

Inclusion Criteria	Exclusion Criteria
1. Able to provide a written informed consent 2. Patients within the age limit (18–65 years) 3. Patients diagnosed with MDD (*DSM-IV*) a. Newly diagnosed (new cases) b. Newly started (old cases): i. Restarted on antidepressants (2-week washout) (31 pax.) ii. Switched to a new antidepressant (3 pax.)	1. Patients having psychotic, cognitive disorders 2. Patients with a history of any other drug abuse 3. Pregnant patients 4. Patients with epilepsy

7.2.3 Clinical Questionnaires

In this study, clinical assessments of the MDD patients were performed based on two clinically relevant questionnaires, that is, the Hospital Anxiety and Depression Scale (HADS) and the Beck Depression Inventory-II (BDI-II). During this process, numeric scores were achieved and were considered as gold standard. For the sake of convenience, translated Malay and English versions of the questionnaires were used (also attached in Appendix A), as the study was conducted in Malaysia and the local language was Malay. The Malay versions of the HADS[5] and BDI-II[6] used were standard clinically proven questionnaires to rate MDD severity. Furthermore, the questionnaires were self-administrated. However, to improve the quality and to maintain the accuracy of the process, the questionnaires were administered under the supervision of an experienced nursing staff.

In this study, a response to treatment was defined as a 50% improvement in clinical symptoms.[7,8] In the literature, the response to SSRI treatment was consistently reported to range from 50% to 60%.[9–13] In this study, the response criterion involved pre- and posttreatment BDI-II scores. Response to treatment was defined as a 50% improvement in clinical symptoms assessed with the BDI-II scores, that is, a 50% improvement in pre- versus posttreatment BDI-II scores.

After the fourth week, the study participants were labeled as respondents (R) or nonrespondents (NR) according to their scores based on the questionnaires: BDI-II and HADS. Response to treatment was defined as a 50% improvement in clinical symptoms.[7,8] According to the BDI-II, a patient's condition was considered normal if the accumulated score ranged between 0 and 10; as mildly depressed for scores ranged between 11 and 20; as moderately depressed between 21 and 30; as severely depressed between 31 and 40; and as extremely severely depressed between 41 and 63. In addition, according to the HADS, cumulative scores greater than 7 are considered abnormal.

7.2.4 Experimental Setup for EEG/ERP Data Acquisition

Fig. 7.1 shows an experimental setup for EEG/ERP data acquisition. The sole purpose of the experiment design was to record a quality dataset as data quality is considered as a core factor to experiments performed with EEG equipment. Factors such as the correct size of EEG cap and proper cap setup are important for maintaining data quality.

Fig. 7.2 shows the hardware setup for EEG data acquisition. The EEG cap is connected to a Brain Master amplifier, which amplifies the EEG signal for reliable recording of data. A computer system runs the Brain Master Discovery software.

Fig. 7.3 shows a snapshot of the home screen of the Brain Master Discovery software used to setup the parameters for EEG/ERP data recordings. The home screen is used when entering information regarding participants such as identification number (ID), name, any comments, and session number. In addition, the home screen is used when setting up to assess more parameters shown in a separate window: the "setup options" window.

FIGURE 7.1 Experimental setup for EEG data acquisition.

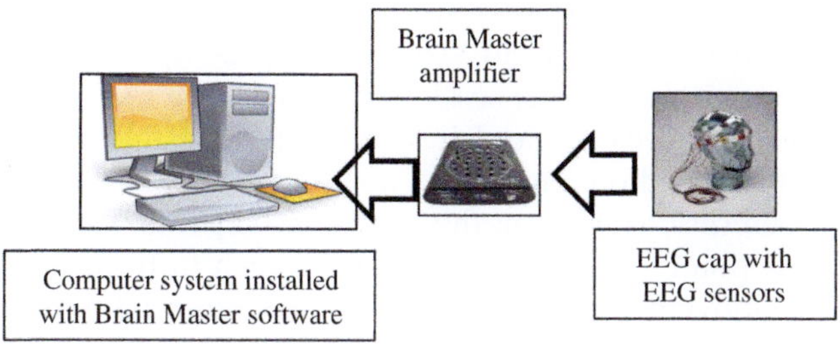

FIGURE 7.2 Block-level representation of EEG/ERP data acquisition.

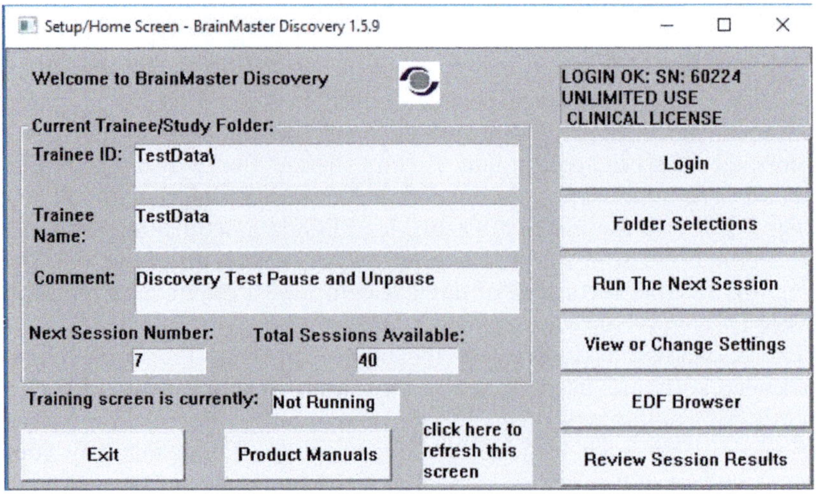

FIGURE 7.3 Home screen for the Brain Master Discovery software.

Fig. 7.4 shows the setup options window of the Brain Master Discovery software. In the setup options window, different parameters can be adjusted according to the experiment design, for example, a new name can be given to the data file. A limited number of channels can be selected, for example, only 8 channels or less can be selected from a list of 19 available channels. In addition, the frequency range of different frequency bands can be adjusted accordingly. This option is especially used in neurofeedback experiments. The training protocol needs to be set to either simulation or training mode. A real EEG data recording requires a training mode setup. Simulation mode helps in recording simulated data only.

Fig. 7.5 shows a flow chart of EEG/ERP recordings divided into four steps, including three experimental physiological conditions. In Step 1, an experimental setup was performed using EEG cap, amplifier, and computer system. An EEG cap suitable for recording the data was selected according to the head measurements (using a measuring tape) of each study participants head size. The cap setup took 20 minutes. Step 2 was an eyes closed (EC) session: patients were required to sit in a chair with their eyes closed and in a maximal possible alert state. EC recordings were based on 5 minutes. Step 3 was for EEG data recording in an eyes open (EO) condition: the study participants were required to look at a fixed point (" + ") on a computer screen in front of them with minimum eye movements and blinks. This task required 5 minutes.

Step 4 was a visual three stimulus oddball task: the study participants were exposed to a random sequence of shapes displayed on a

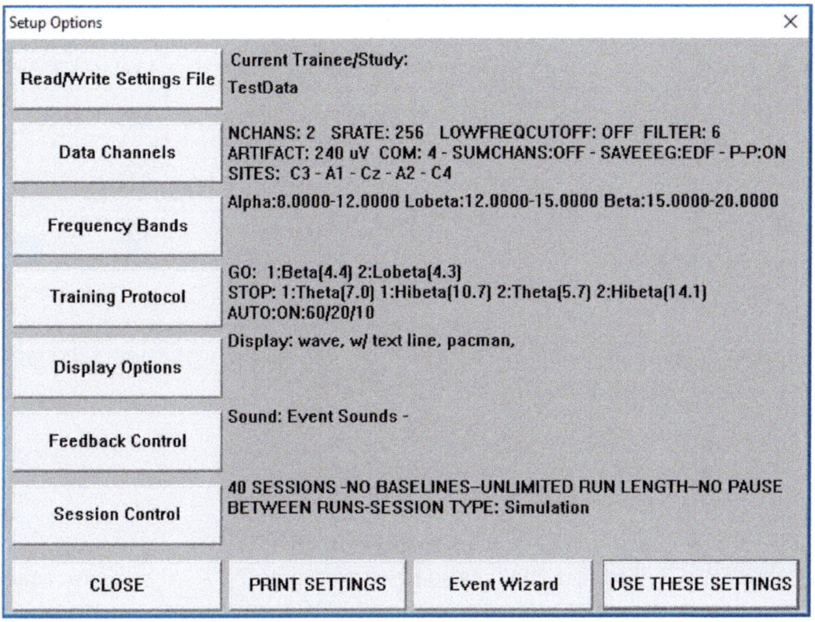

FIGURE 7.4 Setup options for the Brain Master Discovery software.

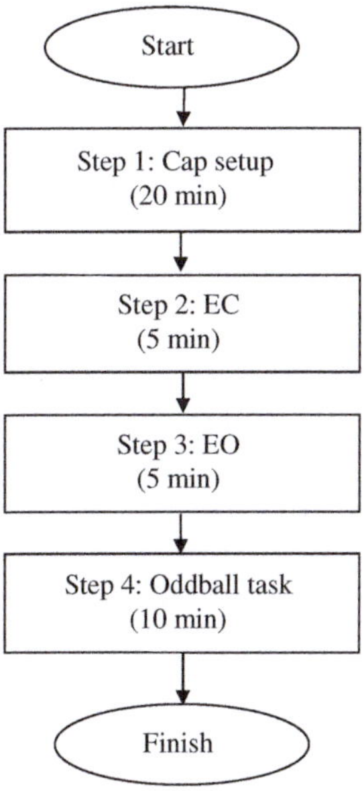

FIGURE 7.5 EEG experiment design.

TABLE 7.3 Visual 3-Stimulus Oddball Task[14]

Stimuli Name	Number of Occurrences	Shape of Stimuli
Standard	314	
Distractor	45	
Target	41	

computer screen in front of them. Only one shape was displayed at a time. There was a total of three shapes (Table 7.3):

1. Target: a blue circle 5.0 cm in size.
2. Standard: a blue circle 4.5 cm in size.
3. Distractor: a checker board 18.0 cm in size.

The study participants had to respond to the Target shape by pressing the SPACE key on the keyboard and take "NO" action for other shapes. The recordings were performed for 10 minutes.

In general, the MDD patients were considered to have low cognitive abilities due to the illness. Such abnormalities were observed with a change in P300 intensity and the occurrence or latency of a P300 peak. For example, depression was associated with a delay in the occurrence of a P300 peak[15] and was found in MDD patients when compared with the healthy controls.[16–18] In addition, decreased P300 intensity in the right hemisphere was observed in MDD patients based on low resolution electromagnetic tomography (LORETA) analysis.[19] Moreover, longer P300 latency was observed in a study involving visually evoked stimuli.[20]

7.3 STUDY PARTICIPANTS INFORMATION

In this study, two distinct groups of participants were recruited: (1) MDD patients and (2) healthy controls. Thirty-four MDD patients, (17 males and 17 females) with a mean age of 40.33 and standard deviation of ± 12.861, meeting international criteria for MDD diagnoses according to *Diagnostic and Statistical Manual-IV* (*DSM-IV*)[21] were recruited. In addition, the second group included 30 healthy participants (21 males and 9 females) with a mean age of 38.277 and standard deviation of ± 15.64, termed as the control group. The recruitment of both groups strictly followed the study inclusion and exclusion criteria, as described previously.

Table 7.4 summarizes demographic information available about the MDD patients. According to the table, the responders (R) and nonresponders (NR) were enlisted separately. This discrimination between responders and nonresponders was based on clinical questionnaires. In this study, the HADS and BDI-II were considered as the gold standard for determining the clinical conditions of the MDD patients.

TABLE 7.4 Clinical Characteristics of the MDD Patients, Mean (± SD)

<table>
<tr><th>Information</th><th>Responders</th><th>Nonresponders</th><th>Total</th></tr>
<tr><td>Age (years)</td><td>40.733 (± 13.0245)</td><td>41.176 (± 12.47)</td><td>40.33 (± 12.861)</td></tr>
<tr><td>Gender (female/male)</td><td>8/8</td><td>9/8</td><td>17/16</td></tr>
<tr><td>Pretreatment BDI</td><td>18.444 (± 7.384)</td><td>22.8235 (± 12.476)</td><td>20.633 (± 8.582)</td></tr>
<tr><td>Pretreatment HADS</td><td>11 (± 1.581)</td><td>10.454 (± 3.297)</td><td>10.727 (± 2.439)</td></tr>
<tr><td>SSRI treatment[a]</td><td>E:9, F:2, S:4, Fl:1</td><td>E:4, F:7, S:4, Fl:2</td><td>E:13, F:9, S:8, Fl:3</td></tr>
<tr><td>Sample size calculation</td><td colspan="3">A group of 34 MDD patients were recruited based on the formula[1,2]:
$n = \frac{P(1-P)\cdot(Z_{1-\alpha/2})^2}{e^2}$

where P is the expected proportion (e.g., expected diagnostic sensitivity), is the error limit which is one half of the desired width of the confidence interval, and $Z_{1-\alpha/2}$ is the standard normal Z-value corresponding to a cumulative probability of $1-\alpha/2$. The investigator must specify the best guess for the proportion that is expected to be found after performing the study.[2] For the research project study, the parameter values were:

• significance $\alpha = .05$ (alpha),
• power of test = 80%, $\beta = .2$ (beta),
• expected diagnostic accuracy $P = 90\%$,[4]
• expected error $e = 10\%$.

$P = 0.90, \alpha = 0.05, e = 0.10, Z_{1-\alpha/2} = 1.96$

$n = \frac{(0.90)(0.10)(1.96)^2}{(0.10)^2} \cong 34$</td></tr>
<tr><td>Inclusion criteria</td><td colspan="3">1. Written informed consent
2. Patients between the age of 18 and 65 years
3. Patients diagnosed with MDD (DSM-IV)
  a. Newly diagnosed (new cases)
  b. Newly started (old cases)
    i. Restarted on antidepressants (1-week washout)
    ii. Switched to new antidepressant</td></tr>
<tr><td>Exclusion criteria</td><td colspan="3">1. Patients having psychotic, cognitive disorders
2. Patients with a history of any other drug abuse
3. Pregnant patients
4. Patients with epilepsy</td></tr>
</table>

[a] *SSRI medication administered: E, escitalopram; F, fluvoxamine; S, sertraline; Fl, fluoxetine.*

7.4 ELECTROENCEPHALOGRAPHY-BASED LOCALIZATION FOR DISEASE PATHOLOGY

EEG and ERP data contain information regarding physiological processes inside the human brain. Abnormal physiological processes due to MDD may show aberrant EEG/ERP patterns. Based on EEG recordings, brain activities were visualized as topographical maps. In addition, sLORETA analyses were performed to construct 3D maps of activations.

7.4.1 Topographic Maps of Activations

Fig. 7.6 shows the locations of 19 EEG sensors for measuring data from the scalp. Information regarding the locations of sensors was required to construct brain topographic maps and the required information was saved in a separate file location.[22]

In this study, brain topographic maps were constructed based on scalp EEG/ERP data in a 2D circular view by performing interpolation (biharmonic spline) on a fine Cartesian grid. On the topographic maps, each location was assigned a color.

Fig. 7.7 shows an example topographic plot. While constructing each topographical map, the value assigned to each scalp location was computed based on statistical tests. The test computed statistics that showed differences of values between two groups at a specific location. According to the test results, the statistically quantified differences

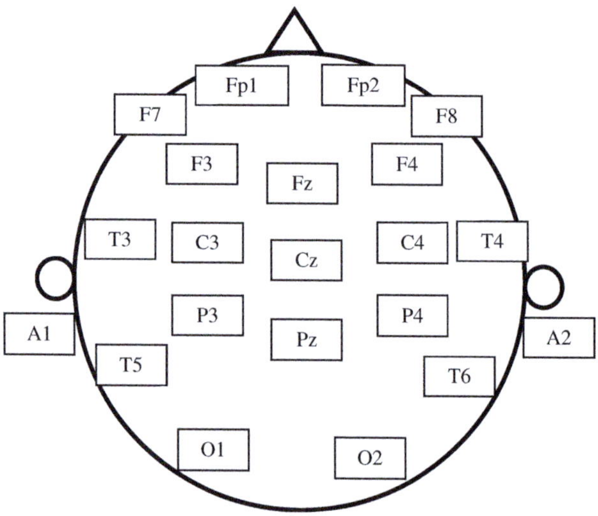

FIGURE 7.6 The 10–20 standard locations of 19 EEG sensors.

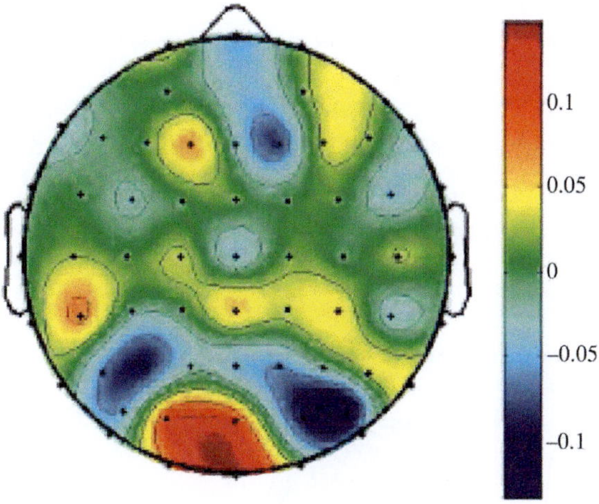

FIGURE 7.7 An example of topographic plot.

between two groups were subjected to color assignments at each scalp location. For example, to determine the differences between the MDD patients and the healthy controls, a *t*-test was performed and the *P*-values were computed. According to a *t*-test, differences between two groups (MDD patients and the healthy controls) are significant if the *P*-value is less than .05. On the other hand, if the *P*-value is greater than .05, the difference is not significant, and the two groups have similar distributions.

The *P*-values were computed at each location and, to produce a visual representation, topographic maps were generated by assigning different colors to each EEG sensor location according to the *P*-values. In this case, topographic maps were constructed based on hypothesis values, that is, $H = 0$ and $H = 1$. Blue was assigned for $H = 0$ and red was assigned where $H = 1$. In other words, red indicated that the identified areas have a statistically significant difference, while blue indicated the absence of significance.

7.4.2 sLORETA Analysis

sLORETA analyses can be used to construct 3D maps of current density due to volume conduction inside the brain while utilizing EEG data.[23] The 3D maps help estimate neuronal sources deep inside the brain. Estimated neuronal sources deep inside the human brain are considered as generators of electric potentials that are observed on the scalp. sLORETA analysis provides for the localization of neuronal sources at different frequencies.

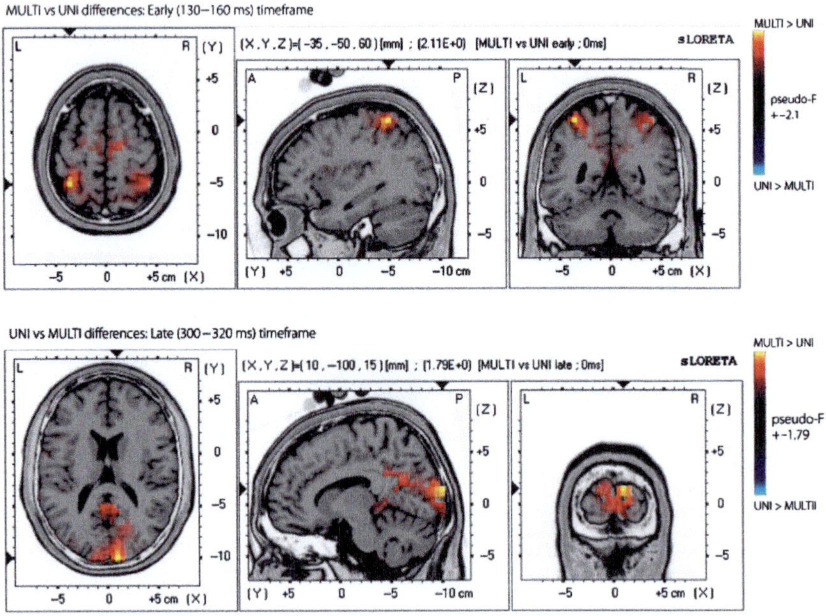

FIGURE 7.8 Example sLORETA plots.

Fig. 7.8 shows an example of topographical plots generated by sLORETA analysis. In this study, 3D maps of MDD patients and the healthy controls were compared in order to analyze differences of these localizations that may illustrate the association of normal or diseased conditions. Similarly, topographic maps were used to investigate the associations between the clinical characteristics of depression and EEG-based topographical maps generated by different EEG features (e.g., power, asymmetry, and coherences).

In comparison to sLORETA analysis, topographical maps of activations were generated with customized EEG-based features. For example, topographic maps were constructed based on coherence, asymmetry, and power. However, in the case of sLORETA analysis, the 3D maps were based on computed current densities only.

Clean EEG data were used to construct 3D topographic maps of activations that show the differences between MDD patients and the controls. Differences between the groups were quantified with distinct colors, for example, red, blue, and gray were used. Let us assume that there were two groups, that is, group A and group B. While constructing a 3D plot, the difference between A and B was assigned a red color if it was positive; a blue color was assigned in cases where a negative value was achieved; and the gray color was used in cases where the difference was equal or zero. The objective of performing a sLORETA analysis was to localize brain areas that showed significant differences between the two groups. This, in turn, will be helpful during the reduction or selection of noteworthy features in EEG data. Brain areas that are affected due to depression were discovered as mentioned in the literature as well. However, the replication of results was performed by EEG data with specific features.

7.5 LOW-DIMENSIONAL REPRESENTATION

In this study, kernelized principal component analysis[24] was used to visualize clustering behavior. Here, Nr-dimensional feature space was reduced to a two-dimensional subset of the feature space, so it could be represented on a two-dimensional plane. This procedure was rotated and nonlinearly transformed the coordinate axes of the feature space to render the view giving the most compact representation. This provided an insight into the clustering and discriminating performance of the feature set and aided in the identification of outliers.

7.6 EEG/ERP DIFFERENCES BETWEEN MAJOR DEPRESSIVE DISORDER PATIENTS AND HEALTHY CONTROLS

In this study, the EEG data showed aberrant patterns for MDD patients, which could not be observed for healthy control subjects. Hence, the differences of EEG (absolute power and interhemispheric asymmetry) and ERP (P300 component) data were computed involving the MDD patients and healthy control groups.

7.6.1 Event-Related Potential Component: P300

In this study, P300 was computed by taking grand averages of segmented EEG data relevant to target shapes. For the MDD patient group, averages were performed over different target trails across all 34 MDD patients. For the control subjects, averages were performed over different target trails across all 30 healthy control subjects. Finally, grand averaged data for the two groups were plotted on the same scales for comparison. The value of P300 peaks and the occurrence times for the two groups were investigated.

7.6.2 Absolute Power and EEG Alpha Interhemispheric Asymmetry

In this study, absolute values were computed for delta, theta, alpha, beta, and gamma bands for the frontal, temporal, occipital, and parietal regions. The values were averaged across different channels in one region. For example, the frontal region included seven electrodes. Therefore, an average value was computed that reflected the overall value of absolute power for the frontal region. Finally, the absolute values were plotted for MDD patients and healthy controls. The final values were compared in order to investigate the increased or decreased values for the respective study group.

7.7 SUMMARY

In this chapter, the design of study protocol has been explained, which has vital importance for the study and to record quality EEG data. Study protocol includes sample size calculation, recruitment criteria, clinical questionnaires, and the procedure for EEG/ERP data acquisition. In addition, the procedure for localization of disease pathology based on topographical maps and sLORETA was elaborated on as well. The computation of P300 peaks and the comparison of absolute power and EEF alpha interhemispheric asymmetry was performed. The purpose was to investigate the significant differences of measure for MDD patients.

References

1. Fosgate GT. Practical sample size calculation for surveillance and diagnostic investigations. *J Vet Diagn Invest*. 2009;21:3–14.
2. Zhou XH, McClish DK, Obuchowski NA. *Statistical Methods in Diagnostic Medicine.* United States of America: John Wiley & Sons; vol. 569, 2009.
3. Khodayari-Rostamabad A, Reilly HP, Hasey GM, et al. A machine learning approach using EEG data to predict response to SSRI treatment for major depressive disorder. *Clin Neurophysiol*. 2013;124(10):1975–1985.
4. Khodayari-Rostamabad A, Hasey GM, MacCrimmon DJ, et al. A pilot study to determine whether machine learning methodologies using pre-treatment electroencephalography can predict the symptomatic response to clozapine therapy. *Clin Neurophysiol.* 2010;121:1998–2006.
5. Yusoff N, Low WY, Yip C-H. Psychometric properties of the Malay version of the hospital anxiety and depression scale: a study of husbands of breast cancer patients in Kuala Lumpur, Malaysia. *Asian Pac J Cancer Prev*. 2011;12:915–917.
6. Mahmud WMRW, Awang A, Herman I, et al. Analysis of the psychometric properties of the Malay version of Beck Depression Inventory II (BDI-II) among postpartum women in Kedah, north west of peninsular Malaysia. *Malays J Med Sci*. 2004;11(2):19.

7. Trivedi MH, Fava M, Wisniewski SR, et al. Evaluation of outcomes with citalopram for depression using measurement-based care in STAR*D: implications for clinical practice. *Am J Psychiatry*. 2006;163(1):28–40.
8. Beck AT, Steer RA, Brown GK. *Manual for the Beck Depression Inventory-II*. San Antonio, TX: Psychological Corporation; 1996.
9. Papakostas GI, Fava M. A metaanalysis of clinical trials comparing moclobemide with selective serotonin reuptake inhibitors for the treatment of major depressive disorder. *Can J Psychiatry*. 2006;51(12):783.
10. Papakostas GI, Thase ME, Fava M, et al. Are antidepressant drugs that combine serotonergic and noradrenergic mechanisms of action more effective than the selective serotonin reuptake inhibitors in treating major depressive disorder? A meta-analysis of studies of newer agents. *Biol Psychiatry*. 2007;62(11):1217–1227.
11. Ruhé HG, Huyser J, Swinkels JA, et al. Switching antidepressants after a first selective serotonin reuptake inhibitor in major depressive disorder: a systematic review. *J Clin Psychiatry*. 2006;67(12):1836–1855.
12. Carvalho AF, Machado JR, Cavalcante JL. Augmentation strategies for treatment-resistant depression. *Curr Opin Psychiatry*. 2009;22(1):7–12.
13. Entsuah AR, Thase ME. Response and remission rates in different subpopulations with major depressive disorder administered venlafaxine, selective serotonin reuptake inhibitors, or placebo. *J Clin Psychiatry*. 2001;62(11):1,478–877.
14. Polich J, Criado JR. Neuropsychology and neuropharmacology of P3a and P3b. *Int J Psychophysiol*. 2006;60(2):172–185.
15. Bruder GE, Towey JP, Stewart JW, et al. Event-related potentials in depression: influence of task, stimulus hemifield and clinical features on P3 latency. *Biol Psychiatry*. 1991;30(3):233–246.
16. Blackwood D, Sharp C, Walker M, et al. Implications of comorbidity for genetic studies of bipolar disorder: P300 and eye tracking as biological markers for illness. *Brit J Psychiatry*. 1996;Supplement(30):85–92.
17. O'donnell B, Vohs J, Hetrick W, et al. Auditory event-related potential abnormalities in bipolar disorder and schizophrenia. *Int J Psychophysiol*. 2004;53(1):45–55.
18. Souza VB, Muir WJ, Walker MT, et al. Auditory P300 event-related potentials and neuropsychological performance in schizophrenia and bipolar affective disorder. *Biol Psychiatry*. 1995;37(5):300–310.
19. Kawasaki T, Tanaka S, Wang J, et al. Abnormalities of P300 cortical current density in unmedicated depressed patients revealed by LORETA analysis of event related potentials. *Psychiatry Clin Neurosci*. 2004;58(1):68–75.
20. Zhu Y, Chen X-S, Qiu J-Y. Study on visual P300 evoked by facial expression stimulus in patients with depression. *J Shanghai Jiaotong Univ*. 2012;10:014.
21. A.P. Association. *Diagnostic and Statistical Manual of Mental Disorders*. DSM-IV-TR®: American Psychiatric Pub; 2000.
22. Delorme A, Makeig S. EEGLAB: an open source toolbox for analysis of single-trial EEG dynamics including independent component analysis. *J Neurosci Methods*. 2004;134(1):9–21.
23. Pascual-Marqui RD. Standardized low-resolution brain electromagnetic tomography (sLORETA): technical details. *Methods Find Exp Clin Pharmacol*. 2002;24(Suppl D):5–12.
24. Muller K-R, Mika S, Ratsch G, et al. *An introduction to kernel-based learning algorithms. Handbook of Neural Network Signal Processing*. CRC Press; 2001.

CHAPTER

8

Electroencephalography-Based Diagnosis of Depression

8.1 INTRODUCTION

This chapter presents an ML scheme—the *Intelligent Treatment Management System (ITMS)*—that can automatically classify a study sample as either a depressed patient or a healthy control. The description involves efficient identification of the most relevant and discriminant features that could be potential candidates for efficient classification. In this context, Fig. 8.1 shows the block-level representation of *ITMS* including the feature selection and classification for unipolar depressed patients. The *ITMS* presented in this chapter inherently involves subprocesses such as noise removal from the electroencephalography (EEG) data, EEG-based feature generation, and selection of the most suitable features along with the supervised classification and validation, including 10-CV. Moreover, the *ITMS* has two versions. First, the *ITMS diagnosis* intends to perform EEG-based diagnoses of depression by identifying the depressed patients and healthy controls recruited for this research study. Second, the *ITMS treatment selection* performs an EEG-based estimation of treatment efficacy for antidepressant therapy while classifying the major depressive disorder (MDD) patients as either treatment respondents (R) or nonrespondents (NR) involving the pretreatment EEG data recorded from the depressed patients in the study.

The scheme for *ITMS diagnosis* involves an integration of EEG-derived features, including absolute power of distinct EEG bands, the EEG alpha interhemispheric asymmetry, and synchronization likelihood (SL). On the other hand, *ITMS treatment selection* involves the integration of features computed from wavelet transform (WT) analysis, including wavelet energy, wavelet-based sample entropy (WSE), wavelet-based composite permutation entropy index (WCPEI), and wavelet-based fractal dimension (WFD). In addition, this chapter elaborates on the

DOI: https://doi.org/10.1016/B978-0-12-817420-3.00008-4

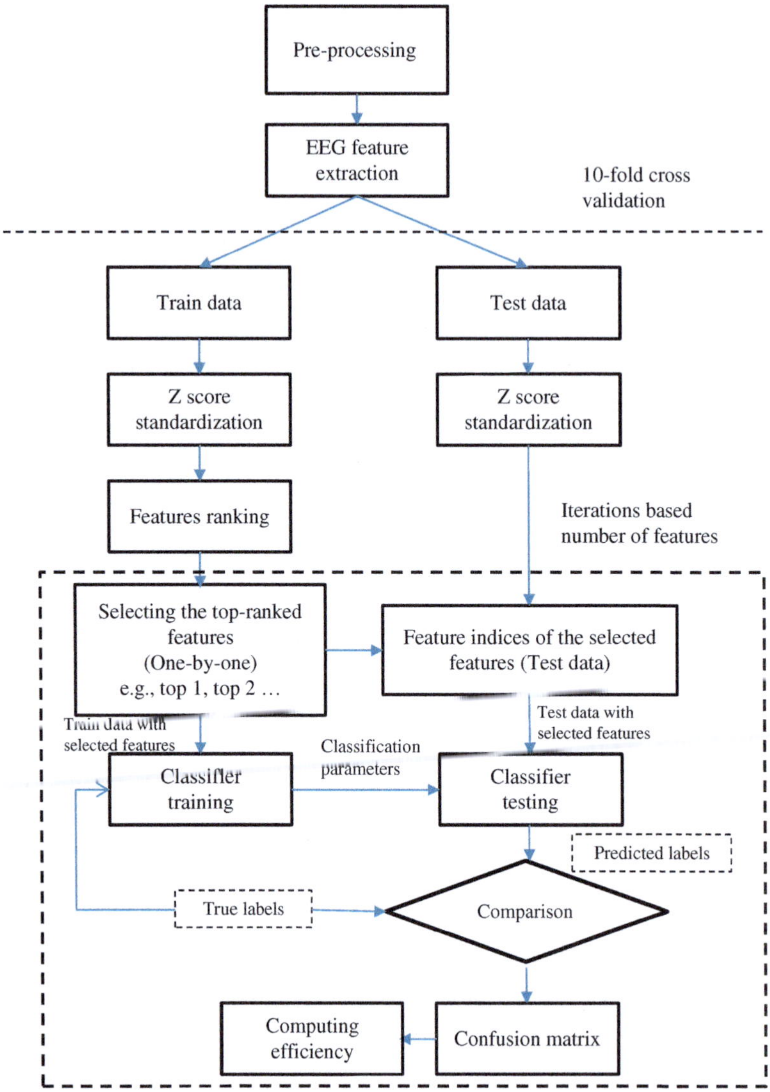

FIGURE 8.1 The *ITMS* with innovative feature selection method.

EEG-based localization of the abnormal brain areas affected due to MDD (as identified in Chapter 2: Electroencephalography Fundamentals). Specifically, this chapter provides the technical details of the *ITMS diagnosis* while Chapter 9, Electroencephalography-Based Treatment Efficacy Assessment Involving Depression, elaborates on the *ITMS treatment selection*.

The *ITMS diagnosis* method offers validation using the EEG data acquired from the depressed patients and healthy controls in the study. The details of the EEG data recording procedure and recruiting study participants are provided in Chapter 7, Design of an Electroencephalography Experiment for Assessing Major Depressive Disorder. The recorded data were termed as the *raw EEG data* and were confounded with background additive noises. Usually, these noises should be removed before performing EEG analysis. Therefore, the removal of noises from the *raw EEG data* is an essential preliminary step, often termed preprocessing. The preprocessing included the correction of EEG data from artifact types such as eye movements (horizontal and vertical), blinks, and muscular and heart activities. See Chapter 7, Design of an Electroencephalography Experiment for Assessing Major Depressive Disorder for a discussion of the details on the EEG preprocessing.

According to Fig. 8.1, the EEG-derived features provide input data to the ML-based method. More importantly, the extracted features should be specific to the target classes that helped identify the most relevant patterns. After feature extraction, the derived features were arranged columnwise in the *EEG data matrix*. The columns represented individual features whereas the rows corresponded to the observations, including each study participant (both the MDD patients and healthy controls). In this study, the features extraction included both eyes closed (EC) and eyes open (EO) conditions and concatenated; hence, the *EEG data matrices* constructed include features corresponding to both the EC and EO datasets.

Because the data comes from different study participants and may require normalization, the *EEG data matrix* underwent *z*-score standardization. In addition, the subsequent statistical methods could be confounded if applied on the data without performing any a priori data standardization or noise reduction. Therefore, data normalization/standardization was necessary before performing any subsequent analysis. The process was repeated for each feature, for example, in the *EEG data matrix*, X_i represented each feature with "i" as a feature index. The *z*-score was computed featurewise by subtracting each value of the feature by the feature mean and divided by the feature standard deviation. A further detail on the *z*-score standardization is provided in the relevant Section 8.4.

In this chapter, the validation of the trained classifier models [logistic regression (LR) classifier, support vector machine (SVM) classifier, and Naïve Bayesian [NB] classifier] required the independence of training and testing sets which was achieved with 10-CV. According to 10-CV, during an iteration the observations (rows) in the *EEG data matrices* were randomly chosen as part of either the training or testing sets. The

10-CV was performed to ensure that the operations such as the data normalization/standardization, feature selection, and classification (Fig. 7.1) were performed separately for the training and testing sets (these processes will be explained in this chapter).

For each fold of the 10-CV, the performance measures such as the classification accuracies, sensitivities, specificities, and the F-measures were computed. The classification models were built independently based only on the training data and the testing was performed separately. Therefore, the computed efficiencies were termed as the test efficiencies (accuracies, sensitivities, specificities, and the *F*-measures). The remainder of this chapter explains the importance and technical details of each subblock of the ML scheme (Fig. 7.1).

8.2 ELECTROENCEPHALOGRAPHY PREPROCESSING

During EEG recordings, noises are usually added due to multiple factors such as power line noise, eye blinking, heartbeats, and yawning. For example, a 50 Hz signal due to power supply line induces noise and is a source of artifact. Eye blinks could be identified with relatively higher amplitude than the normal EEG signal. On the other hand, muscle artifacts could be identified as high frequency bursts. To achieve artifact-free EEG data, the artifacts must be identified and removed while keeping the relevant information intact. However, it is obvious that the rejection of EEG segments with artifacts could also result in the loss of useful information regarding neural activity. Hence, in this study, the correction of the artifacts was preferred over rejection or deletion.

In this chapter, noise removal was performed on 5 minutes of EC and 5 minutes of EO data to achieve at least 2 minutes each of clean EEG data corresponding to EC and EO conditions as per study participant, respectively. A 2 minutes length of clean EEG segments ensured a valid representation of the underlying neural activity.[1]

In Fig. 8.2, a snapshot of the eye blink artifact is shown. In the EEG data, it could be identified by relatively higher amplitudes than a normal EEG activity. The eye blink artifacts normally appear in the frontal sensors, that is, the Fp1 and Fp2, because of their spatial locations near the eyes. As shown in the figure, occurrence of high amplitude involving Fp1 and Fp2 signals causes distortion in the other signals as well (such as F7 and F3, etc.) and corrupts the whole EEG segment. Hence, the eye blink artifacts can be identified easily during a visual inspection of EEG data recordings.

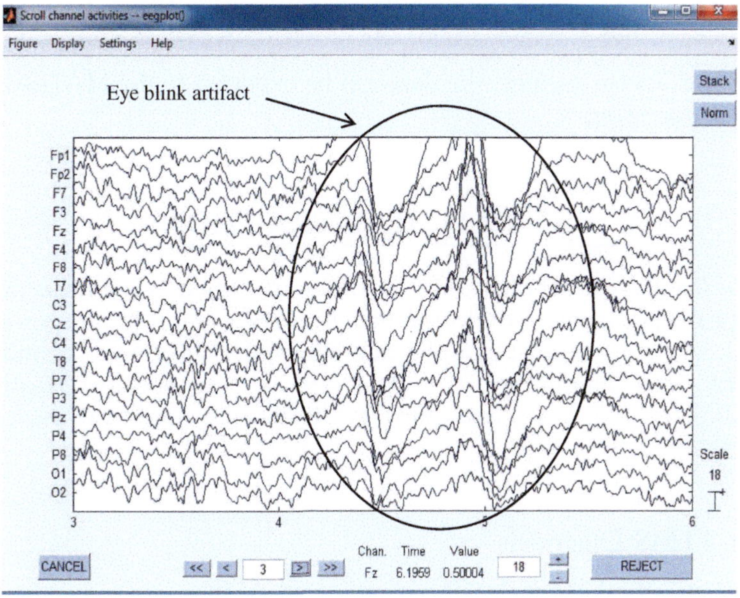

FIGURE 8.2 Example dataset with eye blink artefact.

In this chapter, the artifacts (eye blinks, muscular activities, 50 Hz line noise) were removed using BESA software. The raw EEG data were imported in the European data format (EDF). In BESA, cleaning EEG data from artifact types, namely eyes blinks, muscle activity, line noise, and heart activity, was based on a semi-automatic procedure. First, the researcher should define an artifact type such as eye movements, eye blinks, or muscular and heart activities. Second, the researcher should mark a few portions of the EEG data that were corrupted by the artifact type defined in the previous step so that the software can learn the artifact shape. Based on this learning, the software automatically marks such artifacts in the rest of the EEG recording by automatically traversing the whole recording of the EEG data. As a result, an artifact template was constructed while averaging the marked artifacts. The template corresponded to only one of the artifact types (eye movements, eye blinks, or muscular and heart activities) and could only be learned for one type of artifact at a single time. Furthermore, the noise topography corresponding to the artifact type was automatically estimated and constructed by BESA. Third, the noise topography accompanied with an appropriate user-dependent head model (selected in BESA) were used to autocorrect the artifacts at the marked places in the whole EEG data recording. This technique of artifact correction is called the multiple source eye correction (MSEC) method.[2] Finally, the procedure was repeated by generating the corresponding noise topographies for all possible artifact types observed in the recorded EEG data. After the necessary artifact corrections, the data can be exported to the MATLAB for further processing.

8.3 FEATURE EXTRACTION

The feature extraction results in different *EEG data matrices*. In the case of *ITMS diagnosis*, the computed features were absolute power of each frequency band, EEG alpha interhemispheric asymmetry, and SL. On the other hand, the *ITMS treatment selection* included features, including WCPEI, WSE, and WFD. A detailed description of each feature is provided next.

8.3.1 ITMS Diagnosis

The classification of MDD patients and healthy controls involved EEG spectral power, EEG alpha interhemispheric asymmetry, and SL. Fig. 8.3 shows the feature extraction stage (FES) of *ITMS diagnosis*. The

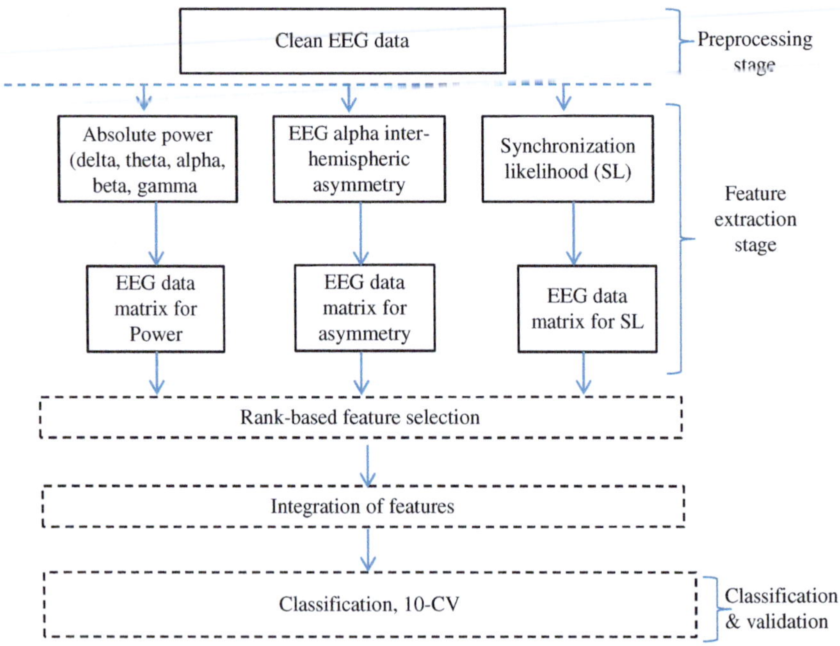

FIGURE 8.3 Feature extraction stage (FES) for *ITMS diagnosis*.

clean EEG data were subjected to the extraction of features. As a result, the EEG data matrices were constructed based on individual feature values and on the integration of absolute power, EEG alpha interhemispheric asymmetry, and SL. Finally, most noteworthy features were selected from the EEG data matrices and were used to perform classification and validation.

8.3.1.1 Significance of the Features (ITMS Diagnosis)

The *ITMS diagnosis* advocates integrating EEG absolute power, EEG interhemispheric asymmetry, and SL. This section provides justifications for this claim. The use of EEG spectral power (EEG activity) as a feature is, in accordance with the literature, termed as the marker for the vulnerability of depression. For example, the EEG activity localized in frontal and temporal areas has been correlated with cognitive deficits and functional impairments that are common characteristics of depression.[3,4] Moreover, according to a recent review, decreased left frontal activity (measured as increased alpha power/amplitude) has been associated with depression.[5]

MDD patients exhibit various abnormalities in the spectral power of different EEG bands. Therefore, this study hypothesized that the analysis of spectral power for MDD and healthy control subjects could result into higher classification results. Furthermore, the association of abnormal EEG activities with the depressed patients could be translated into clinical applications. Hence, the EEG spectral power is the main part for *ITMS diagnosis*.

Numerous studies have shown the significance of EEG alpha interhemispheric asymmetry during depression diagnosis.[6–8] For example, a research study involving the EEG alpha interhemispheric asymmetry has shown association with the psychomotor retardation during depression[6]; another study has declared the EEG frontal asymmetry as a vulnerability marker for depression.[7] In addition, depression has been associated with decreased alpha waves,[8] while another study found that the altered structure of an EEG oscillatory pattern correlated with depression.[5] Hence, EEG alpha interhemispheric asymmetry could be considered as a biomarker for automatic diagnosis of depression.

In addition to the alpha band, activity in other bands such as the theta band has shown relevance during depression such as a decreased frontal theta activity.[9–11] Moreover, depressed patients have exhibited hypoactivation of the left frontal[12,13] and hyperactivation in the right frontal regions.[9] However, the clinical implications of the frontal EEG alpha interhemispheric asymmetry and frontal midline theta activity have been largely unclear.[14] Hence, this situation further warrants an investigation on the applications of EEG features such as EEG alpha interhemispheric asymmetry and EEG spectral power of different frequency bands for diagnosing depression. In addition, fusion of these features could provide more robust features for the ML model for depression diagnosis.

The literature has evidenced that the EEG-based assessment of functional connectivity (FC) between different brain regions could potentially become biomarkers to diagnose depression.[15–18] For example, the aberrant FC between different brain regions have been associated with MDD. In another study, the SL was used to diagnose Alzheimer's disease[19]; however, the SL was not investigated as a feature in a ML-based scenario. Therefore, in this study, the quantification of FC was performed by computing SL values between different sensor pairs at sparse locations.[20] Table 8.1 summarizes EEG features and the significance of absolute power, EEG interhemispheric asymmetry, and the SL. The next section provides a mathematical description of these features.

TABLE 8.1 Features for *ITMS Diagnosis*

EEG features	Significance
Absolute power	The EEG alpha band has shown relevance with MDD patients. Based on the findings, it was hypothesized that these features may show relevance to the proprietary data acquired in this study.
EEG alpha interhemispheric asymmetry	The EEG alpha interhemispheric asymmetry has shown strong association with MDD. Hence, it was hypothesized that the EEG patterns may show a strong association with MDD and ultimately used for classification purposes.
Synchronization likelihood (SL)	Due to the abnormalities reported in the functional connectivity of a depressed brain, it was hypothesized that the SL could be utilized as a feature to classify MDD patients from healthy controls.
Integration of absolute power, EEG alpha interhemispheric asymmetry, and SL	The integration of significant features from individual features was performed based on the hypothesis that the integration of these features will significantly improve the classification accuracy over the accuracy achieved from individual features.

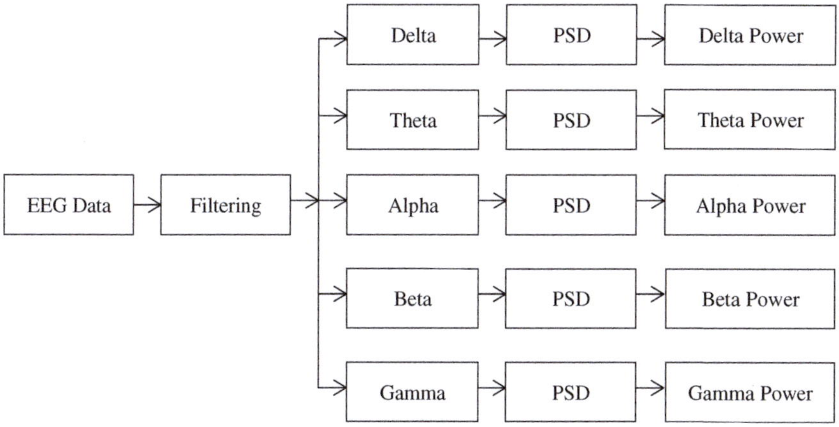

FIGURE 8.4 Filtering and power computation of EEG data.

8.3.1.2 Computing Power for Different Electroencephalography Bands

Fig. 8.4 shows a block-level representation describing the EEG filtering and power spectral density (PSD) computations. The filtering implicates into individual frequency bands such as delta (0.5–4 Hz), theta (4–8 Hz), alpha (8–12 Hz), beta (12–30 Hz), and gamma (30–70 Hz). In this study, the absolute power computation involves 19 EEG channels. The power values are averaged according to the number of channels corresponding to each brain lobe such as frontal (Fp1, Fp2, F3, F4, F7, F8, Fpz), temporal (T3, T4, T5, T6), parietal (P3, P4, P7, P8), occipital (O1, O2), and central (C3, C4).

The EEG signal power was estimated using the Welch periodogram method[21] using the Hanning window function.[22] According to this method, the EEG signal was segmented into eight segments with 50% overlap. The number and overlaps between segments is a variable and adjustments can be done. The PSD was computed for each segment. Finally, an average was performed over all the segments to achieve the absolute power of the observed EEG signal [as described in Eq. (8.1)]. Eq. (8.1) shows the average of the Fourier transform of the windowed EEG signals (periodograms) segmented according to the Welch periodogram method:

$$\hat{S}_{xx}(k) = \frac{1}{N}\left|\sum x(n)e^{-(2\pi jkn/N)}\right|^2 \tag{8.1}$$

where N is the number of segments and n is data samples. Finally, the EEG power in different frequency bands and scalp locations become features during the ML process.

8.3.1.3 Computing Electroencephalography Alpha Interhemispheric Asymmetry

EEG alpha interhemispheric asymmetry shows the difference of power computed between the left and right hemispheres. More specifically, the EEG alpha interhemispheric asymmetry is the difference of spectral power in alpha band.[23] Eqs. (8.2) and (8.3) show the relative EEG signal power in the left and right hemisphere, respectively:

$$W'_{\mathrm{Lmn}} = \frac{\sum_{f=f_1}^{f_2} s_{\mathrm{Lmn}}}{\sum_{f=0.5Hz}^{30Hz} s_{\mathrm{Lmn}}} \tag{8.2}$$

$$W'_{\mathrm{Rmn}} = \frac{\sum_{f=f_1}^{f_2} s_{\mathrm{Rmn}}}{\sum_{f=0.5\ \mathrm{Hz}}^{30\ \mathrm{Hz}} s_{\mathrm{Rmn}}} \tag{8.3}$$

where f_1 and f_2 represent the lower and upper frequency limits of the selected EEG frequency bands, respectively. In addition, the s_{Lmn} and s_{Rmn} implicate the left and right hemispheric spectral power densities, respectively. Furthermore, W_{Lmn} and W_{Rmn} denote the left and right signal power individually.

The computation of the EEG alpha interhemispheric asymmetry involves each study participant for the 16 scalp locations. The odd EEG sensors named Fp1, F7, F3, T3, T5, C3, P3, and O1 represent the left hemisphere. Similarly, the even electrodes termed Fp2, F8, F4, T4, T6, C4, P4, and O2 represent the right hemisphere. Finally, Eq. (8.4) provides the mathematical description of the interhemispheric asymmetry:

$$A_{\mathrm{mn}}(f_1, f_2) = \frac{W'_{\mathrm{Lmn}} - W'_{\mathrm{Rmn}}}{W'_{\mathrm{Lmn}} + W'_{\mathrm{Rmn}}} \times 100 \tag{8.4}$$

where W_{Lmn} and W_{Rmn} represent the left and right signal power, respectively. The EEG alpha interhemispheric asymmetry is denoted by $A_{\mathrm{mn}}(f_1, f_2)$.

The computation of EEG alpha interhemispheric asymmetry involves each channel pair for brain lobes such as the frontal (Fp1, Fp2, F3, F4, F7, F8), temporal (T3, T4, T5, T6), parietal (P3, P4, P7, P8), occipital (O1, O2), and central (C3, C4). Hence, the EEG alpha interhemispheric asymmetry computation for the channel Fp1 involved channel pairs such as Fp1–Fp2; Fp1–F4; Fp1–F8; Fp1–T4; Fp1–T6; Fp1–P4; Fp1–P8; Fp1–O2; and Fp1–C4.

8.3.1.4 Computing Functional Connectivity With Synchronization Likelihood

SL is a FC measure that is normally computed for all possible electrode combinations in an EEG recorded dataset. Since EEG data could have multiple channels, the number of combinations depends on the number of electrodes. This section provides the mathematical representation for computing the SL measure.

The multiple EEG channels are represented as M simultaneous time series $x_{k,i}$, where k implicates number of channels such as $k = 1, 2, 3, \ldots, M$, and i represents time points such as $i = 1, 2, 3, \ldots, N$. In step 1 for computing the SL, the EEG data are used to represent in a high-dimensional space involving the time delay embedding method.[24] As shown in Eq. (8.5), a channel of EEG data is used to construct embedding vectors $X_{k,i}$:

$$X_{k,i} = (x_{k,i}, x_{k,i+l}, x_{k,i+2l}, \ldots, x_{k,i+(m-1)l}) \tag{8.5}$$

where l is lag and m is the embedding dimension.

In step 2 for computing the SL, for each channel k and time point i, a probability $P^{\varepsilon}_{k,i}$ is calculated according to Eq. (8.6). More specifically, a comparison is performed between the difference of the embedding vectors and a threshold value. According to the following equation, the distances less than a threshold value contribute toward the probabilities:

$$P^{\varepsilon}_{k,i} = \frac{1}{2(\omega_2 - \omega_1)} \sum_{\substack{j=1 \\ \omega_1 < |i-j| < \omega_2}}^{N} \theta\left(\varepsilon - |X_{k,i} - X_{k,j}|\right) \tag{8.6}$$

where $|.|$ represents the Euclidean distance and θ represents the Heaviside step function, $\theta(x) = 0$ if $x \le 0$ and $\theta(x) = 1$ for $x > 0$. In addition, ω_1 and ω_2 implicate two windows such that ω_1 is the Theiler correction for autocorrelation effects and should be at least of the order of the autocorrelation time.[25] Moreover, the ω_2 represents a window that sharpens the time resolution of the synchronization measure. Their values are selected such that $\omega_1 << \omega_2 << N$.

In step 3 for computing the SL, once all the probabilities are computed for each time point i, now for each k critical distances $\varepsilon_{k,i}$ are computed such that their probabilities should follow a formula, that is, $P^{\varepsilon}_{k,i} = p_{\text{ref}}$, where $p_{ref} << 1$.

In step 4 for computing the SL, for each time pair (i,j) and within the considered window ($\omega_1 < |i-j| < \omega_2$; e.g., $w_1 = 1$ and $w_2 = 3$), it is easy to count the number of channels $H_{i,j}$ [as shown in Eq. (8.6)], for which the embedding vectors $X_{k,i}$ and $X_{k,j}$ will be closer together considering the critical distance $\varepsilon_{k,i}$:

$$H_{i,j} = \sum_{k=1}^{M} \theta\left(\varepsilon_{k,i} - |X_{k,i} - X_{k,j}|\right) \tag{8.7}$$

Eq. (8.7) represents a count on the number of embedded signals showing a resemblance with each other. This number lies in a range between 0 and M.

In step 5 for computing the SL, for each channel k and each discrete time pair (i,j), the SL is computed as:

$$\text{if}\left|X_{k,i} - X_{k,j}\right| < \varepsilon_{k,i}{:}S_{k,i,j} = \frac{H_{i,j} - 1}{M - 1}$$

$$\text{if}\left|X_{k,i} - X_{k,j}\right| \geq \varepsilon_{k,i}{:}S_{k,i,j} = 0$$

By averaging over all j the SL $S_{k,i}$ is obtained as in Eq. (8.8):

$$S_{k,i} = \frac{1}{2(\omega_2 - \omega_1)} \sum_{\substack{j=1 \\ \omega_1 < |j-i| < \omega_2}}^{N} S_{k,i,j} \tag{8.8}$$

The SL $S_{k,i}$ could be interpreted as a measure describing how strongly channel k at time i is synchronized to all the other $M-1$ channels. The SL may have different values between P_{ref} and 1. Specifically, $S_{k,i} = P_{\text{ref}}$ represents a scenario where all M channels are uncorrelated. On the other hand, the $S_{k,i} = 1$ refers to maximal synchronization between all M channels. Preferably, the P_{ref} should have a low value such that it is independent from the properties of the time series or embedding parameters.

For a 19-channel EEG cap from Brain Master, the SL can be computed involving each channel pair for brain lobes such as the frontal (Fp1, Fp2, F3, F4, F7, F8, Fpz), temporal (T3, T4, T5, T6), parietal (P3, P4, P7, P8), occipital (O1, O2), and central (C3, C4). In particular, the SL for Fp1 involved the channel pairs: Fp1−Fp2; Fp1−F4; Fp1−F8; Fp1−T4; Fp1−T6; Fp1−P4; Fp1−P8; Fp1−O2; and Fp1−C4. Further, the extracted SL features could be arranged columnwise in a matrix, called the *EEG data matrix*. The *EEG data matrix* may have reductant or irrelevant features when subjected to ML classification. Hence, the matrix underwent feature selection.

Here the feature selection method is briefly described (refer to Section 8.5 for a complete description of the feature selection method). This chapter advocates a rank-based feature selection method employed to find the most significant features in the *EEG data matrix*.[26] According to this method, each feature is assigned a weight value corresponding to the ability of a feature to classify the data points into corresponding target classes. The weight values were computed according to the area under curve (AUC) of the receiver operating characteristic (ROC) computed for a feature. A higher weight value was assigned to a feature with larger AUC than a feature with a smaller AUC. The weight value could be any number between 0 and 1, indicating a bad to good classification ability. The weight values allowed listing the most significant features in a list in descending order. According to this method, the most noteworthy features could be identified by selecting the top-listed features only.

8.3.1.5 Integration of Features

In this study, the features are integrated based on a feature concatenation method. In particular, an integration of the features was done by selecting the most discriminant features from EEG power, EEG alpha interhemispheric asymmetry, and the SL.

8.4 STANDARDIZATION

This study advocates the use of *z*-scores standardization. The standardization eliminates possible outliers in the data matrix. In *z*-score standardization, each feature is subjected to standardization separately. In particular, the mean and standard deviation of a feature is computed. Next, the *z*-score could be computed by performing subtraction of each element value by its mean and dividing by its corresponding standard deviation.

Mathematically, the EEG data standardization was performed using the formulas described in Eqs. (8.9) and (8.10).[27]

For example, X_i represents an element of each feature with "*i*" as an index. For a feature, the mean $\overline{X}$ and the standard deviation S are computed, respectively [as shown in Eq. (8.10)] where the N represents total number of elements in a feature. Finally, the *z*-value is computed according to Eq. (8.9).

$$Z = \frac{X_i - \overline{X}}{S} \tag{8.9}$$

$$\overline{X} = \frac{\sum_{i=1}^{N} X_i}{N}, \quad S = \sqrt{\frac{\sum_{i=1}^{N} (x_i - \overline{x})^2}{N-1}} \tag{8.10}$$

8.5 FEATURE SELECTION

The selection of most relevant features resulted in dimension reduction that further facilitates a classifier to learn actual patterns in the data rather than be confounded to any outlier. Hence, the feature selection would certainly improve classifier performance.[28] The selection of most significant features essentially includes two steps[26]: first, assigning a weight to each feature in the *EEG data matrix* based on their classification ability while computing its relevance with the target vector, and second, identification of redundant features based on computing correlations among top-ranked features.

Assigning a weight to each feature was performed according to the criterion, that is, ROCs.[29] According to the criterion, the empirical ROC curve was computed for each feature and the AUC was computed, termed as a z-value. The z-value may vary between 0 and 0.5 describing bad to good classification, respectively. For example, a feature with larger z-value was more capable in discriminating the two classes than a feature with a smaller z-value. Those features with better class separability according to the z-value were ranked higher in the ranking list than the features with lower z-values. As a result, the features were arranged (ranked) in descending order according to the z-values. However, redundancy among distinctive features could not be found just by assigning z-values to the features. In other words, the interrelation among features was not categorized with the feature-ranking method. Therefore, the additional step of computing correlation among features was performed and the features with the highest correlations were discarded. Hence, combining steps 1 and 2, a reduced set of the most relevant and nonredundant features were achieved. The classifier performances were computed for all features while adding one feature at a time. However, only the best classification results were presented. The rest of this section explains the theoretical description of the features' selection method.

Let $A = [a_1, a_2 \ldots a_m]$ be the set of m features and r be a function that assigns a value of merit to each feature $a \in A$. The feature ranking is a function F that assigns a value of merit (relevance) to each attribute $(a_i \in A)$ and returns a list of attributes $(a_i^* \in A)$ observed by its relevance, with $i \in \{1, \ldots, m\}$.

$$F(\{a_1, a_2, \ldots, a_m\}) = < a_1^*, a_2^*, \ldots, a_m^* >$$

where $r(a_1^*) \geq r(a_2^*) \geq \ldots \geq r(a_m^*)$.

By convention, it was assumed that a high score was indicative of a relevant feature. The features were sorted in a decreasing order of ranking for each feature as mentioned by $r(a^*)$. ROCs were defined as the ranking criterion for individual features independently to the context of others. The selection of features was restricted to supervised learning because the information about the treatment outcome was provided based on the clinical scores.

For each feature, the given examples were sorted in descending order according to the feature values. Let r_t be the rank of the tth class -1 examples. Let q_t be the number of class $+1$ examples whose ranks are higher than r_t. The ROC curve is generated by plotting the (t_i, q_t) over

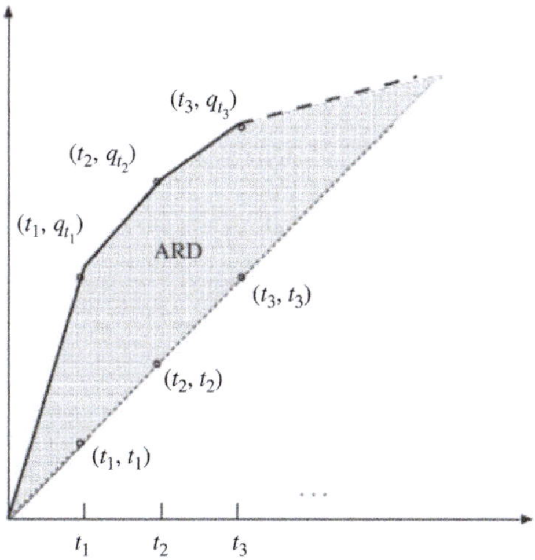

FIGURE 8.5 Receiver operating characteristics curve.[26]

$t=1 \ldots M_{-1}$. Fig. 8.5 shows the ROC curve generated by plotting (t_i,q_t) over $t=1, \ldots, M_{-1}$.

As shown in the figure, the area ARD is defined as the area between the ROC curve and the diagonal line. The ARD was obtained by first computing the distance between the point (t_i,q_t) of the ROC curve and a point (t_i,t_i) on the diagonal line, that is, $q_t - t_i$, and sum up $q_t - t_i$ over $t=1 \ldots M_{-1}$ as follows:

$$\begin{aligned} \left|\sum_t (q_t - t_i)\right| &= \left|\sum_t (r_t - 2t_i)\right| \\ &= \left|\sum_t r_t - 2\sum_t t_i\right| \\ &= \left|\Omega_i(-1) - M_{-1}(M_{-1}+1)\right| \end{aligned} \tag{8.11}$$

where $\Omega_i(-1)$ is the sum of the ranks of class -1 examples when the sorting was performed for all examples by the values of feature i. Eq. (8.12) depends on the size of M_{-1}, and so by dividing it by the whole area M_{+1} and M_{-1}, the ARD of the feature i is described as:

$$\text{ARD}_i = \left|\frac{\Omega_i(-1) - M_{-1}(M_{-1}+1)}{M_{+1}M_{-1}}\right| \tag{8.12}$$

The pseudocode for the feature-ranking method has been provided in Table 8.2.

8.5.1 Example 1

Let a variable (or a set) be defined by capital X and a value of that variable by a same small letter, that is, x. Let C be a set of target classes

TABLE 8.2 Pseudocode for the Feature Ranking Method

```
patterns = [x y];
patterns = sortrows(patterns,-1);
y = patterns(:,2);
p = cumsum(y==1);
tp = p/sum(y==1);
n = cumsum(y==-1);
fp = n/sum(y==-1);
n = length(tp);
Y=(tp(2:n)+tp(1:n-1))/2;
X = fp(2:n) - fp(1:n-1);
auc=sum(Y.*X)-0.5;
```

TABLE 8.3 Real EEG Data

Sample ID	...	*i*	*j*	...	Label
1		−0.2	+0.5		(−)
2		−1.4	−1.4		(−)
3		+0.8	−0.9		(−)
4		−0.8	+0.2		(+)
5	...	+0.1	−2.5	...	(+)
6		+0.5	+1.4		(−)
7		+1.6	−0.3		(+)
8		−2.1	−1.2		(−)
9		−0.3	+2.2		(+)
10		+3.4	−1.7		(−)

or a variable corresponding to the target label. Let F be a variable corresponding to the entire feature set, taking values f. Let F_i be a variable corresponding to ith features taking values f_i.

Table 8.3 enlists the EEG data. According to this description, the features are enlisted columnwise. The table shows two features with 10 example points. In the rest of this section, the computation of the area under the ROC curve is provided.

Tables 8.4–8.6 lists the intermediate values of different variables during the computation of the AUC for the given example data listed in Table 8.3.

TABLE 8.4 Computation for Example 1

Feature Values (Sorted in Descending Order)	Label	p	n	tp	fp
3.4	(−)	0	1	0	0.1667
1.6	(+)	1	1	0.25	0.1667
0.8	(−)	1	2	0.25	0.333
0.5	(−)	1	3	0.25	0.5
0.1	(+)	2	3	0.5	0.5
− 0.2	(−)	2	4	0.5	0.6667
− 0.3	(+)	3	4	0.75	0.6667
− 0.8	(+)	4	4	1	0.6667
− 1.4	(−)	4	5	1	0.8333
− 2.1	(−)	4	6	1	1

TABLE 8.5 Computation of $Y = (\text{tp}(2{:}n) + \text{tp}(1{:}n-1))/2$

tp(2:n)	0.25	0.25	0.25	0.5	0.5	0.75	1	1	1
tp(1:n−1)	0	0.25	0.25	0.25	0.5	0.5	0.75	1	1
(tp(2:n) + tp(1:n−1))/2	0.25	0.5	0.5	0.75	1	1.25	1.75	2	2
Y	0.125	0.25	0.25	0.375	0.5	0.625	0.875	1	1

TABLE 8.6 Computation of $X = (\text{fp}(2{:}n) - \text{fp}(1{:}n-1))$

fp(2:n)	0.1667	0.333	0.5	0.5	0.6667	0.6667	0.6667	0.8333	1
fp(1:n−1)	0.1667	0.1667	0.333	0.5	0.5	0.6667	0.6667	0.667	0.8333
X	0	0.1667	0.1667	0	0.1667	0	0	0.1667	0.1667

AUC = sum($Y.\times X$)−0.5;

AUC = (0.125 × 0 + 0.25 × 0.1667 + 0.25 × 0.1667 + 0.375 × 0 + 0.5 × 0.1667 + 0.625 × 0 + 0.825 × 0 + 0.1667 × 1 + 0.1667 × 1) − 0.5

AUC = 0.5 − 0.2 = 0.3

8.5.2 Example 2

Tables 8.7–8.9 enlist the intermediate values of different variables during the computation of the AUC for the given example data listed in Table 8.3.

TABLE 8.7 Computations for Example 2

Feature Values (Sorted in Descending Order)	Label	p	n	tp	fp
+ 2.2	(+)	1	0	0.25	0
+ 1.4	(−)	1	1	0.25	0.1667
+ 0.5	(−)	1	2	0.25	0.3333
+ 0.2	(+)	2	2	0.50	0.3333
− 0.3	(+)	3	2	0.75	0.3333
− 0.9	(−)	3	3	0.75	0.5
− 1.2	(−)	3	4	0.75	0.6667
− 1.4	(−)	3	5	0.75	0.8333
− 1.7	(−)	3	6	0.75	0.8333
− 2.5	(+)	4	6	1	1

TABLE 8.8 Computation of $Y = (\text{tp}(2{:}n) + \text{tp}(1{:}n-1))/2$

tp(2:n)	0.25	0.25	0.50	0.75	0.75	0.75	0.75	0.75	1
tp(1:$n-1$)	0.25	0.25	0.25	0.50	0.75	0.75	0.75	0.75	0.75
(tp(2:n) + tp(1:$n-1$))/2	0.25	0.25	0.375	0.625	0.75	0.75	0.75	0.75	0.875
Y	0.125	0.125	0.1875	0.3125	0.375	0.375	0.375	0.375	0.4375

TABLE 8.9 Computation of $X = (\text{fp}(2{:}n) - \text{fp}(1{:}n-1))$

fp(2:n)	0.1667	0.333	0.333	0.333	0.5	0.6667	0.8333	0.8333	1
fp(1:$n-1$)	0	0.1667	0.333	0.333	0.333	0.5	0.6667	0.8333	0.8333
X	0.1667	0.1667	0	0	0.1667	0	0.1667	0	0.1667

$\text{AUC} = \text{sum}(Y. \times X) - 0.5;$

$\text{AUC} = (0.125 \times 0.1667 + 0.125 \times 0.1667 + 0.1875 \times 0.1667 + 0.3125 \times 0 + 0.375 \times 0 + 0.375 \times 0.1667 + 0.375 \times 0 + 0.375 \times 0.1667 + 0.4375 \times 0.1667) - 0.5$

$AUC = 0.0833$

8.6 CLASSIFICATION MODELS

8.6.1 Logistic Regression Classification

This study advocates the use of LR classification because it is suitable for solving binomial classification problems. In addition, the LR classifier is capable of finding a suitable model between categorical and continuous variables. For example, the categorical variables are depressed versus normal or treatment respondent versus nonrespondents. The features extracted from the EEG lie in the category of the continuous variables. For example, the LR classification has been commonly applied for various epidemiological studies, for example, during cancer classification either as malignant or benign.[30]

In this study, the objective of the classification process is to model the relationship between significant QEEG (Quantitative EEG) features such as the wavelet-based features and the treatment outcome, that is, respondents (R) versus nonrespondents (NR). In the case of diagnosis, the model was based on a relationship between QEEG features and the classes, that is, MDD patients and healthy controls. The learned model was further utilized to diagnose the MDD patients and to predict antidepressant's treatment outcome.

A multivariate relationship between the EEG-based features and the clinical outcomes, that is, R and NR, was modeled based on the LR model, which is commonly applied for various epidemiological studies (e.g., during classification of cancer either as malignant or benign).[30] In case of diagnosis, the clinical outcomes were the "MDD patients" and "healthy controls." The reduced set of EEG features was considered as the independent variables and the corresponding treatment outcomes (R or NR) were the dependent variables. Logistic function provides the mathematical base on which the logistic model is based and is given by Eq. (8.13):

$$F(z) = E\left(\frac{Y}{x}\right) = \frac{1}{1+e^{-z}} \tag{8.13}$$

where Y was the class labels and assigned a value either (R Vs NR) or (MDD patients Vs healthy controls) and x represents a combination of the EEG features after feature selection, that is, the coefficients achieved by WT technique and the features extracted from EMD and STFT analysis. To obtain the LR model from the logistic function, we used Eq. (8.14):

$$z = \alpha + \beta_1 X_1 + \beta_2 X_2 + \ldots + \beta_k X_k \tag{8.14}$$

where z is a linear combination of α plus β_1 multiplied with X_1, plus β_2 multiplied with X_2, and plus β_k multiplied with X_k, where the X_k are the independent variables and α and β_i are constant terms representing unknown parameters. Furthermore, by replacing the value of z from Eqs. (8.13) to (8.14), the following Eq. (8.15) represents the logistic function:

$$F(z) = E\left(\frac{Y}{x}\right) = \frac{1}{1+e^{-(\alpha+\sum \beta_i X_i)}} \tag{8.15}$$

In terms of response and nonresponse, the risk of a person to be nonresponder or a responder is estimated and represented by Y or $l(x)$. The LR classifier resulted into a likelihood value $l(x)$, where $0 \leq l(x) \leq 1$, which was an indication of subjects associated either with the R or NR category. If $l(x)$ was greater than the *threshold* = 0.5, the subject was declared as R (responder) and otherwise as a NR (nonresponder). In summary, the LR classifier generated probability values to cater for MDD patients as either R or NR to the treatment. The fitting of the LR model is explained as follows.

Assume that we have a sample of n independent observations (x_i, y_i), $i = 1, 2, \ldots, n$, where y_i denotes the value of a dichotomous outcome variable and x_i is the value of the independent variable for the ith subject. The outcome variable is coded as R or NR, $y = [\text{R}, \text{NR}]$, based on the clinical results provided by the BDI-II and HADS. This coding for a dichotomous outcome is used throughout the text. As in the univariate case, fitting the model requires that we obtain estimates of vectors $\beta' = (\beta_0, \beta_1, \ldots, \beta_p)$. The method of estimation used in the multivariable case will be the same as the univariate situation maximum likelihood. In general, the maximum likelihood yields values for the unknown parameters which maximize the probability of obtaining the observed set of data. To apply this method, we must first construct a function, called the likelihood function. This function expresses the probability of the observed data as a function of unknown parameters. The maximum likelihood estimators of these parameters are chosen to be those values that maximize this function. Thus, the estimators are those which agree most closely with the observed data. The likelihood function is nearly identical to that given in Eq. (8.16):

$$l(\beta) = \prod_{i=1}^{n} \pi(x_i)^{y_i}[1 - \pi(x_i)]^{1-y_i} \tag{8.16}$$

8.6.2 Support Vector Machine Classification

The SVM is as a high-efficiency classification model. This study utilizes SVM classifier with linear kernel. The SVM classifies the feature space based on a hyperplane that separates MDD patients and controls according to class labels.[31] The SVM works well for a high-dimensional dataset by establishing a linear decision boundary. In the context of small datasets, using linear kernel SVM instead of a nonlinear kernel reduces the risk of overfitting. In addition, it improves the classification performance by significantly reducing the model complexity. In summary, the LR classifier involved probability values to classify MDD patients versus controls. On the other hand, the SVM concluded a hyperplane to achieve the classification between the two groups. Following paragraph provides a brief description of SVM, mathematically.

If the training data is linearly separable, then a pair (w,b) exists such as $w^T x_i + b \geq 1$ for all $x_i \varepsilon P$ and $w^T x_i + b \leq -1$ for all $x_i \varepsilon P$ with the decision rule given by $f_{w,b}(x) = sgn(w^T x_i + b)$, where w is termed as weighted vector and b as the bias. It is easy to show that when it is possible to linearly separate two classes an optimum separating hyperplane can be found by minimizing the squared norm of the separating hyperplane. The minimization can be set up as a convex quadratic programming (QP) problem: Minimize $\Phi(w) = \frac{1}{2}\|w\|^2$ subject to $y(wx + b) \geq 1$, $i = 1, 2, \ldots, l$.

In the case of linearly separable data, once the optimum separating hyperplane is found, the data points that lie on its margin are known as support vector points and the solution is represented as a linear combination of only these points.

8.6.3 Naïve Bayesian Classification

The NB classification[32] generates conditional posterior probabilities for each data sample. Because the NB classifier is a supervised classifier, it involves the target groups, that is, MDD patients and healthy controls. In case of a binomial classification, the NB classifier produces two probabilities. The classifier model assigns the sample to a class having higher posterior probability.

The NB classifier has a simple structure. NB networks are composed of directed acyclic graphs with only one parent (representing the unobserved node) and several children (corresponding to observed nodes). There is a strong assumption of independence among child nodes in the context of their parent.[33] Thus, the independence model (NB) is based on estimating[34] [Eq. (8.17)]:

$$R = \frac{P(i|X)}{P(j|X)} = \frac{P(i)P(X|i)}{P(j)P(X|j)} = \frac{P(i)\prod P(X_r|i)}{P(j)\prod P(X_r|j)} \tag{8.17}$$

where P indicates probability, i and j are labels, and X contains the predictors or independent variables. Comparing these two probabilities, the larger probability indicates the class label to be more likely an actual class label (if $R > 1$ predict i else predict j).

8.7 VALIDATION

This study introduces 10-CV as its validation method. The 10-CV aims to determine the robustness of the classification models.[35] It allows

TABLE 8.10 Confusion Matrix

	Condition positive	**Condition negative**
Test outcome positive	True positive	False positive
Test outcome negative	False negative	True negative

computing the performance metrics such as the classifier accuracies, sensitivities, and specificities. In this study, the cross-validation provides sampling of data points in the EEG data matrix in such a way that each point can be utilized during classification testing and training.

In brief, the 10-CV includes the division of data samples into 10 subgroups. The classifier training involves nine of the data samples and the last one employed during testing the classification model. The division of training and testing samples are subject to random distribution. This randomness successfully removes the bias of classification models to individual data points. Table 8.10 represents an example of a confusion matrix for the binary classification problem. The matrix provided the quantification of performances of the classification models employed in this study.

Eqs. (8.18)–(8.22) provide mathematical descriptions of the performance metrics computed from the confusion matrix. By definition, the sensitivity of a classification model is defined as the percentage of true cases (TP) that are correctly classified as cases as shown in Eq. (8.18). For example, a depressed subject is classified as depressed by the classification model. Moreover, the classifier specificity refers to the percentage of true noncases (TN) correctly classified as noncases. Eq. (8.19) shows the mathematical description. For example, a classification model predicts a study sample as a healthy control who is actually a healthy control, according to the ground truth. Furthermore, the classification accuracy illustrates the percentage of correctly classified cases and noncases among all the example points. Eq. (8.20) provides the mathematical description. In addition, the false positive (FP) and false negative (FN) corresponds to the incorrect classification of healthy controls as MDD patients and incorrect classification of MDD patients as healthy controls, respectively. Finally, Eqs. (8.21) and (8.22) show the formulas for the positive predictive value (PPV) and the negative predictive values (NPV), respectively.

$$\text{Sensitivity} = \frac{\text{TP}}{\text{TP} + \text{FN}} \tag{8.18}$$

$$\text{Specificity} = \frac{\text{TP}}{\text{FP} + \text{TN}} \tag{8.19}$$

$$\text{Accuracy} = \frac{\text{TP} + \text{TN}}{\text{TP} + \text{TN} + \text{FN} + \text{FP}} \tag{8.20}$$

$$\text{PPV} = \frac{\text{TP}}{\text{TP} + \text{FP}} \tag{8.21}$$

$$\text{NPV} = \frac{\text{TN}}{\text{FN} + \text{TN}} \tag{8.22}$$

By definition, the *F*-measure is defined as a weighted harmonic average of recall and precision values.[36] The *precision* is the probability that a (randomly selected) patient analyzed to be MDD was really MDD, according to the ground truth. On the contrary, the *recall* is the probability that a (randomly selected) MDD patient correctly identified as MDD. Eq. (8.23) shows the F-measure mathematically:

$$F\text{-score} = \frac{2\text{TP}}{2\text{TP} + \text{FP} + \text{FN}} \tag{8.23}$$

where TP refers to true positives, FP refers to false positive, and FN represents false negatives.

8.8 MDD PATIENTS VERSUS HEALTHY CONTROLS

In this section, the differences between the MDD patients and healthy controls are studied involving the sLORETA analysis, topographic maps, comparison of power in different band powers, the alpha interhemispheric asymmetry, and the ERP component P300. The details on each result are provided in the respective sections.

8.8.1 The sLORETA Analysis

Table 8.11 shows sLORETA maps for the MDD patients and healthy controls. During EC condition, the MDD patients have shown increased activity at regions (frontal: BA = 11/47; $T = 7.752$, $P < .05$). On the other hand, during EO conditions regions, the brain activity was manifested in areas (frontal: BA = 47; temporal: BA = 21; occipital: BA = 19, $P < .05$, $T = 6.637$).

The sLORETA analysis for MDD patients and healthy controls implicated the brain regions such as frontal, temporal, and occipital. In literature, numerous studies have reported abnormalities in the frontal cortex associated with MDD. For example, a reduction in prefrontal cortex volume was reported in recent literature.[37,38] In some older studies, ischemic lesions were located in the anterior frontal cortex and found to be associated with more severe depression.[39,40] In addition, patients with

TABLE 8.11 The sLORETA Maps of MDD Patients' Versus Healthy Controls

Student's *t*-test-based differences between MDD patients and healthy controls	**Brain areas**
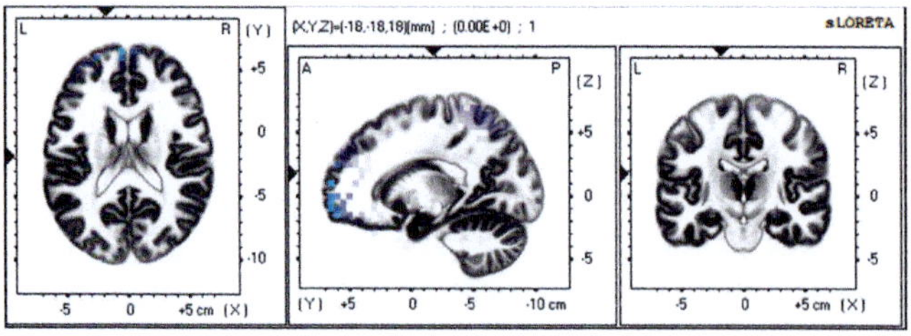 (Eyes closed condition)	Frontal: BA = 11/47, BA = 9/10
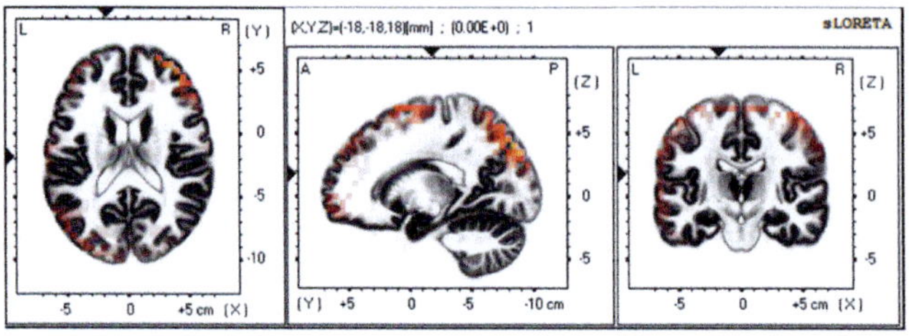 (Eyes open condition)	Frontal: BA = 11; temporal: BA = 21; occipital: BA = 19

ischemic stroke showed a strong correlation between lesions and subsequent depression. The stroke has affected the prefronto-subcortical circuits, particularly in the left hemisphere.[41] Cognitive disability was commonly observed in MDD patients, particularly involving frontal cortex. Its volume reductions were reported to range from 7% overall reduction[42] to 48% in the subgenual prefrontal cortex.[43] In a postmortem study, significant differences between the control subjects and depressed patients were found in prefrontal cortical areas.[44]

In temporal areas, several studies assessed left and right temporal volumes separately.[45–47] Only Vythilingam[47] found evidence for a lateralization effect, reporting smaller left temporal lobe volume in patients. Notably, the patient sample studied by these authors had the longest illness duration compared to other studies.[47] In this regard, the left-lateralized temporal lobe changes may reflect progression of the disease over time or a distinct pathophysiological process that affects risk of relapse. In the occipital area, an increase in theta and alpha activity has been found in MDD patients.[48] In Table 8.12, the brain regions implicated in this section are presented with respect to different frequency bands, including physiological conditions.

TABLE 8.12 Abnormal Brain Regions for MDD Patients

Physiological condition	Brain regions
Eyes closed (EC)	Frontal
Eyes open (EO)	Frontal, temporal, and occipital

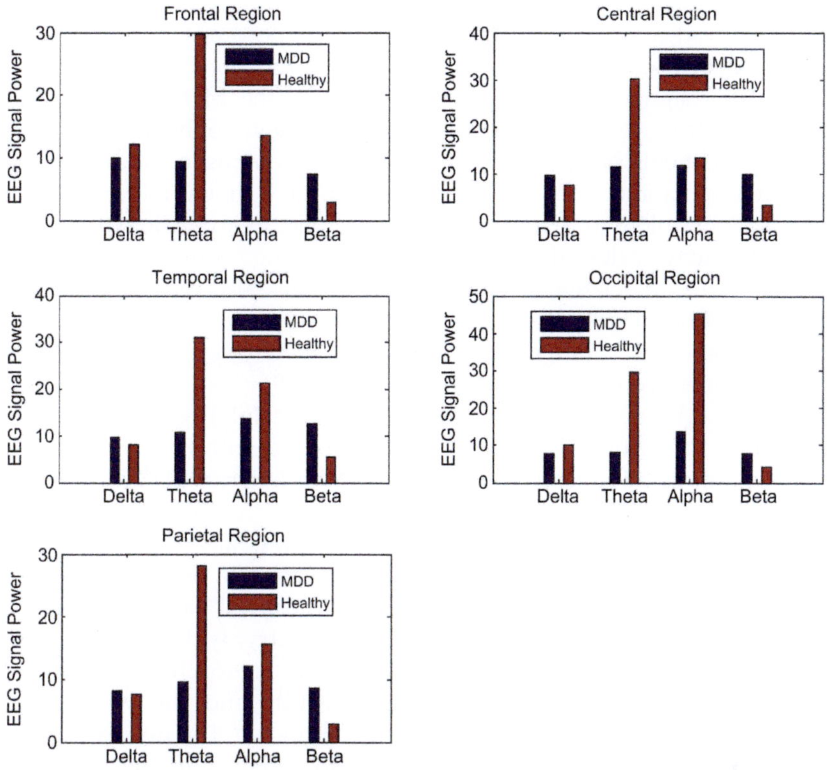

FIGURE 8.6 Eyes open EEG signal power for MDD versus healthy controls ($P < .01$).

8.8.2 Electroencephalography Signal Power and Alpha Interhemispheric Asymmetry

Fig. 8.6 shows EO EEG signal power differences between MDD patients and healthy controls, including frontal, temporal, parietal, occipital, and central regions. The MDD patients exhibited less theta and alpha signal powers in all regions compared with healthy controls. In addition, the frontal and occipital regions have shown less delta signal power in depressed patients compared to healthy controls. However, for brain

TABLE 8.13 EEG Interhemispheric Alpha Asymmetry for MDD Patients Versus Healthy Controls ($P < .01$)

Brain Regions	MDD Patients	Healthy Controls
Frontal	Right < left	Left < right
Parietal, central, temporal, occipital	Left < right	Right < left

regions such as the central, temporal, and parietal areas the delta and beta bands showed a slight increase in EEG signal powers when compared between depressed and healthy controls ($P < .01$).

The EEG signal power differences between the MDD patients and healthy controls implicated less power most commonly found in the depressed patients only. The EEG signal is of a composite nature and the power computations in each frequency band have shown relevance with the pathophysiology of depression. For example, in an earlier study, elevated EEG activity (less alpha) was observed during resting condition[49–52] and an increased relative power (less alpha) was also reported.[53,54] The increase in EEG activity (less alpha value) was observed in frontal, parietal,[55] and occipital[56] brain areas. Moreover, the early stages of depression were characterized by elevated alpha activity.[48]

In Table 8.13, depressed individuals tend to exhibit relatively greater right frontal activity (less alpha) when compared with healthy controls. In short, depressed individuals showed greater anterior EEG activity.

EEG alpha interhemispheric asymmetry has been studied as a vulnerability marker for depression. In 1983, Davidson et al.[57] reported a relative hyperactivation of the right prefrontal cortex. In a later study, the researchers considered "approach" and "withdrawal" as orthogonal to each other and as fundamental to EEG asymmetry, which may become a vulnerability measure for depression.[58] The approach system facilitated appetitive behavior with positive results and the withdrawal system motivated aversive and negative emotions.[59] The decreased left-sided frontal EEG activation was related to a deficit in the approach system. Based on these observations, it was concluded that the study subjects who have such symptoms were at risk of negative emotional states and depression in response to environmental stress.

8.8.3 ERP Component: P300

Fig. 8.7 shows plots of grand averaged P300 components at central region (Cz) that could significantly discriminate the groups: healthy controls and MDD patients. As evident from the figure, the healthy controls exhibited higher P300 intensities than MDD patients. In addition, the MDD patients depicted longer latencies than the healthy controls.

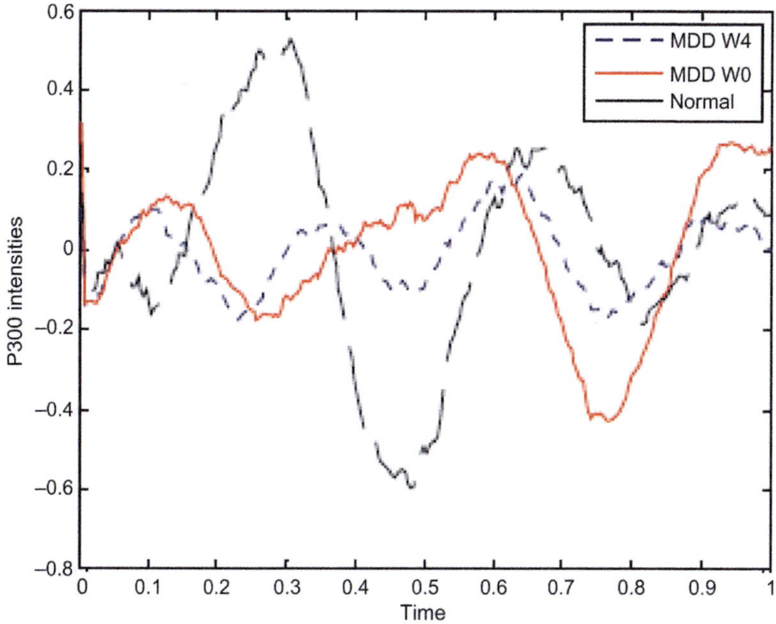

FIGURE 8.7 The P300 for the MDD patients and healthy controls.

Results shown in Fig. 8.7 implicated a decreased P300 intensity and a larger latency for depressed patients. However, the normal behavior is different and can be seen in that the P300 peak occur with 300 ms delay with prominent peak intensity. In the literature, the P300 was associated with cognitive abilities and commonly studied for MDD. For example, the auditory evoked potentials (AEP) have shown positive correlation with cognitive abilities.[60,61]

In general, the MDD patients were considered to have low cognitive abilities due to their illness. Such abnormalities were observed with a change in P300 intensity and occurrence or latency of the P300 peak. For example, depression was associated with a delay in occurrence of P300 peak[62] and only found in MDD patients when compared with the healthy controls.[63–65] In addition, a decreased P300 intensity in the right hemisphere was observed based on low resolution electromagnetic tomography (LORETA) analysis.[66] Moreover, longer P300 latency was observed in a study involving visually evoked stimuli.[67]

8.8.4 Classification Results (ITMS Diagnosis)

Fig. 8.8 shows the diagnosis accuracy as a function of a number of features. The figure shows a direct relationship between the classification

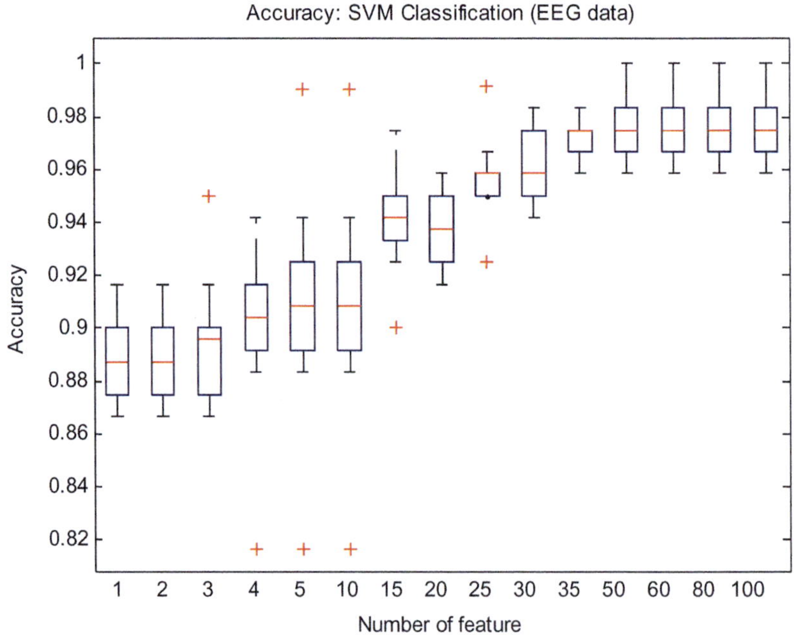

FIGURE 8.8 SVM accuracy as a function of number of features.

accuracy and the number of features. More particularly, an increase in the number of features increases the classification accuracy for SVM. However, at a certain point (i.e., number of features = 35), a further increase in the number of features could not increase the classification accuracy rather it remians constant. It means that only the first 35 features could be sufficient for achieving maximum classification accuracy.

Table 8.14 shows results of the classification performance based on the SVM classification model. According to the table, however, the highest efficiencies were achieved involving integration of features based on the method, that is, *ITMS diagnosis*. Hence, it resulted in accuracy = 98.8%, sensitivity = 98.6%, specificity = 99.4%, and *f*-measure = 98.3. The second highest performance (accuracy = 97.1%, sensitivity = 97%, and specificity = 98%) was achieved with alpha asymmetry. The third highest accuracy was achieved with alpha power features. The alpha power features resulted in accuracy = 95.1%, sensitivity = 96.6%, and specificity = 94.4%.

The study shows high performance accuracies (i.e., 98.8%) which are comparable to previous studies, for example, a different study based on the classification of depressed patients versus healthy controls concluded 90% accuracy.[68] In this context, Knott et al.[11] found 91.3% classification

TABLE 8.14 SVM Classification for MDD Patients' Versus Healthy Controls

EEG-Based Features	Accuracy	Sensitivity	Specificity	*F*-measure
Delta power	84.5	82	87.4	81.4
Theta power	87.5	88.6	87	85.6
Alpha power	95.1	96.6	94.3	94.4
Beta power	90.1	90.3	92.4	89.7
Alpha asymmetry	97.1	97	98	96.2
Synchronization likelihood (SL)	94.38	97	92.4	92.1
ITMS diagnosis	**98.8**	**98.6**	**99.4**	**98.3**

accuracy involving 70 depressed patients and 23 normal subjects. Their study employed linear features. However, the number of samples were not balanced between the two classes. Due to this imbalance between the numbers of samples in two groups, the classification results may be biased toward the depressed patient's class. Moreover, Lee et al.[69] analyzed EEG data involving the detrended fluctuation analysis (DFA) based on 11 depressed and 11 healthy subjects as controls. The study concluded higher DFA values in MDD patients when compared with controls. However, due to the small sample size, the results could not be generalized. In another study, EEG analysis based on wavelet entropy analysis achieved 80% accuracy involving 26 MDD participants only.[70]

In comparison, the ML scheme has achieved the highest classification results while employing linear EEG features only. The results implicate that conventional ML techniques with linear features as input data can achieve high performances. In addition, integrating features from different frequency bands could increase the classification efficiency than the individual frequency bands. This observation is in accordance to the research ideology that useful features may be discovered from EEG subbands.[71] Moreover, the SVM exhibits the highest classification performances by using comparatively more features compared to LR and NB. Because SVM has a complex structure it; therefore requires more data samples to appropriately train compared with LR and NB classifiers.

To remove the likelihood that the resulting classifier models are concluded due to noise present in the EEG data, the following precautions are adopted. First, during preprocessing, the artifacts are carefully removed and tested by plotting their histograms plots. Second, equal sample sizes are selected in both MDD patients and healthy controls. For example, the gender distribution is equal between the groups to eliminate any gender bias from the conclusive results. Third, as overfitting may happen, 100-time iterations of the 10-CV improves the robustness of the underlying classification models.

8.9 SUMMARY

This chapter explains in detail how a ML approach can improve the diagnosis of treatment efficacy assessment for depression. The process of feature extraction, selection, and classification is discussed. The validation of the ML model is necessary, especially for the case of a small study sample. The resulting model can be used to generalize the results. This model could be implemented in a clinical setting and requires only a small amount of EEG data informing on patients' condition and could assist clinicians for possible diagnosis as well.

References

1. Thatcher RW, North DM, Biver CJ. Development of cortical connections as measured by EEG coherence and phase delays. *Hum Brain Mapp*. 2008;29:1400–1415.
2. Berg P, Scherg M. A multiple source approach to the correction of eye artifacts. *Electroencephalogr Clin Neurophysiol*. 1994;90:229–241.
3. McIntyre RS, Cha DS, Soczynska JK, et al. Cognitive deficits and functional outcomes in major depressive disorder: determinants, substrates, and treatment interventions. *Depress Anxiety*. 2013;30:515–527.
4. Jaeger J, Berns S, Uzelac S, Davis-Conway S. Neurocognitive deficits and disability in major depressive disorder. *Psychiatry Res*. 2006;145:39–48.
5. Fingelkurts AA, Fingelkurts AA. Altered structure of dynamic electroencephalogram oscillatory pattern in major depression. *Biol Psychiatry*. 2015;77:1050–1060.
6. Cantisani A, Koenig T, Horn H, Müller T, Strik W, Walther S. Psychomotor retardation is linked to frontal alpha asymmetry in major depression. *J Affect Disord*. 2015;188:167–172.
7. Allen JJ, Reznik SJ. Frontal EEG asymmetry as a promising marker of depression vulnerability: summary and methodological considerations. *Curr Opin Psychol*. 2015;4:93–97.
8. Kan D, Lee P. Decrease alpha waves in depression: an electroencephalogram (EEG) study. In: *International Conference on BioSignal Analysis, Processing and Systems (ICBAPS), 2015*. Kuala Lumpur, Malaysia; 2015:156–161.
9. Saletu B, Anderer P, Saletu-Zyhlarz G. EEG topography and tomography (LORETA) in diagnosis and pharmacotherapy of depression. *Clin EEG Neurosci*. 2010;41:203–210.
10. Coutin-Churchman P, Moreno R. Intracranial current density (LORETA) differences in QEEG frequency bands between depressed and non-depressed alcoholic patients. *Clin Neurophysiol*. 2008;119:948–958.
11. Knott V, Mahoney C, Kennedy S, Evans K. EEG power, frequency, asymmetry and coherence in male depression. *Psychiatry Res. Neuroimaging*. 2001;106:123–140.
12. Henriques JB, Davidson RJ. Left frontal hypoactivation in depression. *J Abnorm Psychol*. 1991;100:535.
13. Kemp A, Griffiths K, Felmingham K, et al. Disorder specificity despite comorbidity: resting EEG alpha asymmetry in major depressive disorder and post-traumatic stress disorder. *Biol Psychol*. 2010;85:350–354.
14. Gold C, Fachner J, Erkkila J. Validity and reliability of electroencephalographic frontal alpha asymmetry and frontal midline theta as biomarkers for depression. *Scand J Psychol*. 2013;54:118–126.

15. Anand A, Li Y, Wang Y, et al. Activity and connectivity of brain mood regulating circuit in depression: a functional magnetic resonance study. *Biol Psychiatry*. 2005;57:1079–1088.
16. Bae JN, MacFall JR, Krishnan KRR, Payne ME, Steffens DC, Taylor WD. Dorsolateral prefrontal cortex and anterior cingulate cortex white matter alterations in late-life depression. *Biol Psychiatry*. 2006;60:1356–1363.
17. Fingelkurts AA, Fingelkurts AA, Rytsälä H, Suominen K, Isometsä E, Kähkönen S. Impaired functional connectivity at EEG alpha and theta frequency bands in major depression. *Hum Brain Mapp*. 2007;28:247–261.
18. Hamilton JP, Gotlib IH. Neural substrates of increased memory sensitivity for negative stimuli in major depression. *Biol Psychiatry*. 2008;63:1155–1162.
19. Stam C, Montez T, Jones B, et al. Disturbed fluctuations of resting state EEG synchronization in Alzheimer's disease. *Clin Neurophysiol*. 2005;116:708–715.
20. Stam C, Van Dijk B. Synchronization likelihood: an unbiased measure of generalized synchronization in multivariate data sets. *Physica D*. 2002;163:236–251.
21. Welch P. The use of fast Fourier transform for the estimation of power spectra: a method based on time averaging over short, modified periodograms. *IEEE T Audio Electroacoust*. 1967;70–73.
22. Oppenheim AV, Schafer RW, Buck JR. *Discrete-Time Signal Processing*. vol. 2. Englewood Cliffs: Prentice-Hall; 1989.
23. Hinrikus H, Suhhova A, Bachmann M, et al. Electroencephalographic spectral asymmetry index for detection of depression. *Med Biol Eng Comput*. 2009;47:1291–1299.
24. Takens F. *Detecting strange attractors in turbulence*. Dynamical Systems and Turbulence, Warwick 1980. Springer; 1981:366–381.
25. Theiler J. Spurious dimension from correlation algorithms applied to limited time-series data. *Phys Rev A*. 1986;34:2427.
26. Mamitsuka H. Selecting features in microarray classification using ROC curves. *Pattern Recognit*. 2006;39:2393–2404.
27. Cheadle C, Vawter MP, Freed WJ, Becker KG. Analysis of microarray data using Z score transformation. *J Mol Diagn*. 2003;5:73–81.
28. Liu H, Motoda H. *Feature Selection for Knowledge Discovery and Data Mining*. vol. 454. Springer Science & Business Media; 2012.
29. Liu H, Motoda H. *Computational Methods of Feature Selection*. CRC Press; 2007.
30. Shen L, Tan EC. Dimension reduction-based penalized logistic regression for cancer classification using microarray data. *IEEE/ACM Trans Comput Biol Bioinform*. 2005;2:166–175.
31. Vapnik VN, Vapnik V. *Statistical Learning Theory*. vol. 2. New York: Wiley; 1998.
32. Mitchell TM. *Machine Learning. WCB*. Boston, MA: McGraw-Hill; 1997.
33. Good IJ. Probability and theWeighing of Evidence, ed: JSTOR; 1950.
34. Nilsson NJ. Learning machines; 1965.
35. Golland P, Liang F, Mukherjee S, Panchenko D. *Permutation tests for classification. Learning Theory*. Springer; 2005:501–515.
36. Van Rijsbergen CJ. *The Geometry of Information Retrieval*. Cambridge University Press; 2004.
37. Salvadore G, Nugent AC, Lemaitre H, et al. Prefrontal cortical abnormalities in currently depressed versus currently remitted patients with major depressive disorder. *NeuroImage*. 2011;54:2643–2651.
38. Treadway MT, Waskom ML, Dillon DG, et al. Illness progression, recent stress, and morphometry of hippocampal subfields and medial prefrontal cortex in major depression. *Biol Psychiatry*. 2014.

39. Robinson RG, Kubos KL, Starr LB, Rao K, Price TR. Mood changes in stroke patients: relationship to lesion location. *Compr Psychiatry*. 1983;24:555–566.
40. Lipsey JR, Robinson R, Pearlson G, Rao K, Price T. Mood change following bilateral hemisphere brain injury. *Br J Psychiatry*. 1983;143:266–273.
41. Vataja R, Pohjasvaara T, Leppävuori A, et al. Magnetic resonance imaging correlates of depression after ischemic stroke. *Arch Gen Psychiatry*. 2001;58:925–931.
42. Coffey C, Wilkinson W, Parashos L, et al. Quantitative cerebral anatomy of the aging human brain A cross-sectional study using magnetic resonance imaging. *Neurology*. 1992;42:527.
43. Drevets WC, Price JL, Simpson JR, et al. Subgenual prefrontal cortex abnormalities in mood disorders. *Nature*. 1997;386:824–827.
44. Rajkowska G, Miguel-Hidalgo JJ, Wei J, et al. Morphometric evidence for neuronal and glial prefrontal cell pathology in major depression. *Biol Psychiatry*. 1999;45:1085–1098.
45. Bremner JD, Vythilingam M, Vermetten E, et al. Reduced volume of orbitofrontal cortex in major depression. *Biol Psychiatry*. 2002;51:273–279.
46. Caetano SC, Hatch JP, Brambilla P, et al. Anatomical MRI study of hippocampus and amygdala in patients with current and remitted major depression. *Psychiatry Res. Neuroimaging*. 2004;132:141–147.
47. Vythilingam M, Vermetten E, Anderson GM, et al. Hippocampal volume, memory, and cortisol status in major depressive disorder: effects of treatment. *Biol Psychiatry*. 2004;56:101–112.
48. Grin-Yatsenko VA, Baas I, Ponomarev VA, Kropotov JD. Independent component approach to the analysis of EEG recordings at early stages of depressive disorders. *Clin Neurophysiol*. 2010;121:281–289.
49. Lemere F. The significance of individual differences in the Berger rhythm. *Brain: J Neurol*. 1936.
50. Begić D, Popović-Knapić V, Grubišin J, et al. Quantitative electroencephalography in schizophrenia and depression. *Psychiatr Danub*. 2011;23:355–362.
51. Jaworska N, Blier P, Fusee W, Knott V. Alpha power, alpha asymmetry and anterior cingulate cortex activity in depressed males and females. *J Psychiatr Res*. 2012;46:1483–1491.
52. Roemer RA, Shagass C, Dubin W, Jaffe R, Siegal L. Quantitative EEG in elderly depressives. *Brain Topogr*. 1992;4:285–290.
53. John ER, Prichep L, Fridman J, Easton P. Neurometrics: computer-assisted differential diagnosis of brain dysfunctions. *Science*. 1988;239:162–169.
54. Prichep LS, John E. QEEG profiles of psychiatric disorders. *Brain Topogr*. 1992;4:249–257.
55. Grin-Yatsenko VA, Baas I, Ponomarev VA, Kropotov JD. EEG power spectra at early stages of depressive disorders. *J Clin Neurophysiol*. 2009;26:401–406.
56. Bruder GE, Sedoruk JP, Stewart JW, McGrath PJ, Quitkin FM, Tenke CE. Electroencephalographic alpha measures predict therapeutic response to a selective serotonin reuptake inhibitor antidepressant: pre-and post-treatment findings. *Biol Psychiatry*. 2008;63:1171–1177.
57. Schaffer CE, Davidson RJ, Saron C. Frontal and parietal electroencephalogram asymmetry in depressed and nondepressed subjects. *Biol Psychiatry*. 1983;18(7):753–762.
58. Henriques JB, Davidson, RJ. Regional brain electrical asymmetries discriminate between previously depressed and healthy control subjects. *J Abnorm Psychol*. 1990;99:22–31.
59. Davidson RJ. Anterior electrophysiological asymmetries, emotion, and depression: conceptual and methodological conundrums. *Psychophysiology*. 1998;35:607–614.

60. Mulert C, Pogarell O, Juckel G, et al. The neural basis of the P300 potential. *Eur Arch Psychiatry Clin Neurosci.* 2004;254:190–198.
61. Volpe U, Mucci A, Bucci P, Merlotti E, Galderisi S, Maj M. The cortical generators of P3a and P3b: a LORETA study. *Brain Res Bull.* 2007;73:220–230.
62. Bruder GE, Towey JP, Stewart JW, Friedman D, Tenke C, Quitkin FM. Event-related potentials in depression: influence of task, stimulus hemifield and clinical features on P3 latency. *Biol Psychiatry.* 1991;30:233–246.
63. Blackwood D, Sharp C, Walker M, Doody G, Glabus M, Muir W. Implications of comorbidity for genetic studies of bipolar disorder: P300 and eye tracking as biological markers for illness. *Br J Psychiatry.* 1996;(Supplement)85–92.
64. O'donnell B, Vohs J, Hetrick W, Carroll C, Shekhar A. Auditory event-related potential abnormalities in bipolar disorder and schizophrenia. *Int J Psychophysiol.* 2004;53:45–55.
65. Souza VB, Muir WJ, Walker MT, et al. Auditory P300 event-related potentials and neuropsychological performance in schizophrenia and bipolar affective disorder. *Biol Psychiatry.* 1995;37:300–310.
66. Kawasaki T, Tanaka S, Wang J, Hokama H, Hiramatsu K. Abnormalities of P300 cortical current density in unmedicated depressed patients revealed by LORETA analysis of event related potentials. *Psychiatry Clin Neurosci.* 2004;58:68–75.
67. Zhu Y, Chen X-S, Qiu J-Y. Study on visual P300 evoked by facial expression stimulus in patients with depression. *J Shanghai Jiaotong Univ.* 2012;10:014.
68. Hosseinifard B, Moradi MH, Rostami R. Classifying depression patients and normal subjects using machine learning techniques and nonlinear features from EEG signal. *Comput Methods Programs Biomed.* 2013;109:339–345.
69. Lee J-S, Yang B-H, Lee J-H, Choi J-H, Choi I-G, Kim S-B. Detrended fluctuation analysis of resting EEG in depressed outpatients and healthy controls. *Clin Neurophysiol.* 2007;118:2489–2496.
70. Li Y, Li Y, Tong S, Tang Y, Zhu Y. More normal EEGs of depression patients during mental arithmetic than rest. In: *Noninvasive Functional Source Imaging of the Brain and Heart and the International Conference on Functional Biomedical Imaging, 2007. NFSI-ICFBI 2007. Joint Meeting of the 6th International Symposium on.* Hangzhou, China; 2007:165–168.
71. Ahmadlou M, Adeli H, Adeli A. Fractality analysis of frontal brain in major depressive disorder. *Int J Psychophysiol.* 2012;85:206–211.

CHAPTER

9

Electroencephalography-Based Treatment Efficacy Assessment Involving Depression

9.1 INTRODUCTION

The previous chapter elaborated on an electroencephalography (EEG)-based machine learning (ML) scheme designed to diagnose and perform treatment efficacy assessment for depression based on EEG data only, known as the *Intelligent Treatment Management System (ITMS)* for depression. The ITMS is common for diagnosis and treatment selection, but not for the feature extraction block. Therefore, this chapter focuses on explaining the feature extraction subblock for *ITMS-treatment selection*. *ITMS-treatment selection* involves an integration of features such as EEG signal energy, wavelet-based sample entropy (WSE), wavelet-based composite permutation entropy index (WCPEI), and wavelet-based fractal dimension (WFD).

Wavelet transformation provides a multiresolution decomposition of EEG data. As a result, different EEG bands are achieved. Furthermore, the decomposed version of the EEG signal is subject to time-domain features to compute signal complexity, including the methods sample entropy (SE), composite permutation entropy (CPEI), and fractal dimension (FD). This chapter elaborates on the importance of these methods in the context of depression. The integration of multiple decompositions of EEG data and time-based features provides insights at multiple levels of EEG data.

ITMS-treatment selection inherently involves supervised classification. Therefore, its validation should involve clinical evidences such as clinically validated scores on disease severity. In this study, the clinical evidences included the administration of questionnaires such as the Beck Depression Inventory (BDI) and Hospital Anxiety and Depression

EEG-Based Experiment Design for Major Depressive Disorder
DOI: https://doi.org/10.1016/B978-0-12-817420-3.00009-6

(HADS) scores. Hence, this study has considered these clinical questionnaires as a gold standard during EEG-based analysis and classification.

Furthermore, this chapter includes EEG-based localizations of abnormal brain areas through the use of sLORETA analysis and topographic plots. The chapter presents the results of ITMS classifications for treatment response.

9.1.1 ITMS-Treatment Selection

This chapter advocates the use of time-frequency decomposition of EEG data based on wavelet transform (WT) analysis to classify treatment respondents and nonrespondents. Fig. 9.1 shows a block-level representation of the feature extraction stage (FES) for ITMS-treatment selection.

9.1.1.1 Significance of Features (ITMS-Treatment Selection)

This chapter advocates for the integration of time and frequency domain features, including the decomposition of EEG into different frequency bands, which could be useful for the treatment efficacy assessment of depression. In addition, as EEG data have high temporal resolution, complex EEG data are computed with the use of techniques such as CPEI, sample entropy (SE), and FD.

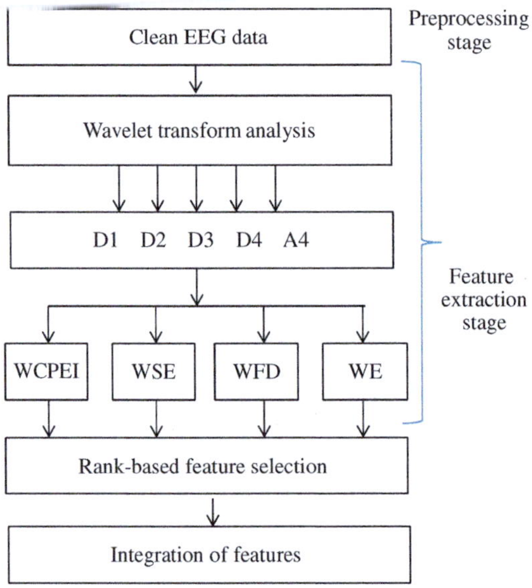

FIGURE 9.1 Feature extraction stage (FES) for ITMS-treatment selection.

During WT analysis, the decomposition of EEG signals results in individual frequency bands, also termed as multiresolution decomposition. In the literature, this multiresolution decomposition has shown important applications as a diagnostic tool for various medical applications, including both signal and image processing.[1,2] Other examples include various biomedical applications,[3,4] including the diagnosis of epilepsy and Alzheimer's disease.[5,6] However, WT analysis could provide meaningful insights by decomposing EEG data at multiple levels of decompositions. Each level of decomposition refers to a particular EEG frequency band. In addition, similar techniques, such as short-time Fourier transform (STFT) and empirical model decomposition (EMD), can provide signal decomposition at multiple levels. Therefore, for the sake of comparison, the study provides EEG data analysis involving both STFT and EMD analyses.

As mentioned in Chapter 4, Pathophysiology of Depression, abnormalities in different EEG frequency bands could result due to different conditions of MDD. For example, increased delta and theta power over the right hemisphere were associated with the MDD patients only.[7] Moreover, LORETA-based studies have revealed elevated delta, theta, and beta power in the brain region, ACC, in MDD patients.[8] In addition, left frontal alpha activity has been reported in depressed patients when compared with controls associated with the left frontal[9] and DLPFC brain regions.[10] Hence, multiresolution decomposition of EEG data into various bands could result in increased chances of discovering MDD treatment respondents or nonrespondents. Also, the prediction of treatment efficacy might be improved, which could help psychiatrists in the selection of a suitable antidepressant for a given MDD patient.

In this study, EEG data were subjected to WT analysis which resulted in various wavelet coefficients denoted as D1, D2, D3, D4, and A4. For example, D1 refers to detailed coefficients at the first level of wavelet decomposition. In addition, A4 refers to approximate coefficients at the fourth level of wavelet decomposition. During this process, the selection of an appropriate window function was based on a criterion know as energy to permutation entropy ratio (EPER). For this criterion, a wavelet window function that provided the highest value of the ratio should be selected for WT analysis. High values of EPER indicated that the wavelet window function showed relevance with the changes in the recorded EEG data. A mathematical description of this criterion is presented in Section 9.1.1.2.

CPEI was used to measure the depth of clinical anesthesia.[11] It has shown promise as a practical EEG measure of GABAergic hypnotic drug effects. For example, it successfully tracks the quantitative assessment of EEG patterns from awake to light to deep anesthesia.

Moreover, it requires minimum preprocessing and is highly resistant to eye blink artifacts. CPEI is a function of frequencies in the EEG signal; for example, it shows high values near to one in cases where EEG signals are dominant at the higher frequency bands. On the contrary, it results in low values, nearly 0.4, in case where EEG data are dominated by the delta and theta bands.

SE represents the signal complexity of the bioelectric processes in the brain; for example, SE values for patients with cerebral injuries are found to be lower than normal subjects.[12] Low values of SE indicate more self-similarity in the time series EEG signal.[13] In another sleep study, the SE values characterized the depth of sleep. The deeper the sleep, the smaller the SE![14]

FD is used to quantify the complexity and self-similarity of an object.[15] FD provides the ability to distinguish between different pathophysiological states by directly examining EEG in time domain. Since the dimension of a plane is equal to 2 and the dimension of a line is equal to 1, it can be expected that the EEG FD will always be between 1 and 2. FD can be computed using different techniques, but Higuchi's algorithm was used here.[16]

In this chapter, time and frequency information are integrated by decomposing EEG signals into various frequency bands and then computing the complexity for each frequency band. For example, EEG decomposition with WT analysis resulted in delta, theta, alpha, and beta frequency bands. Table 9.1 describes complexity measures such as WCPEI, WSE, and WFD. These measures were utilized to compute complexity values corresponding to individual frequency bands. Sections 9.1.1.4–9.1.1.7 will describe their specific details.

9.1.1.2 Selection of an Appropriate Basis Function for Electroencephalography Analysis

The selection of the most suitable wavelet window function to perform the decomposition of the EEG signal has vital importance. In this study, the most suitable wavelet function was selected based on a criterion known as EPER, according to the formula described in Eq. (9.1):

$$S = \frac{\text{Signal}_{\text{E}}}{\text{Signal}_{\text{PE}}} \tag{9.1}$$

where $\text{Signal}_{\text{PE}}$ denotes EEG signal permutation entropy (PE) and Signal_{E} refers to energy. According to the formula, the "*S*" values were computed for all study participants and a final value was selected based on averaging across all study participants. A wavelet window function which provides the highest value of "*S*" was considered as the most suitable window function for wavelet analysis and was selected for

TABLE 9.1 EEG Features for *ITMS-Treatment Selection*

EEG Features	Significance
Wavelet detailed and approximate coefficients	Wavelet coefficients represent the EEG signal at various scales with time information and are believed to be used to successfully reconstruct the EEG signal.
Wavelet-based signal energy (WE)	Signal energy extracted via wavelet coefficients is believed to represent the underlying EEG signal more concisely.
Wavelet-based sample entropy (WSE)	The multiresolution decomposition of the EEG signal is provided by WT analysis. The SE computed the complexity of the wavelet coefficients.
Wavelet-based composite permutation entropy index (WCPEI)	The multiresolution decomposition of the EEG signal is provided by WT analysis. Time-based features such as composite permutation entropy index (CPEI) computed the complexity of the wavelet coefficients.
Wavelet-based fractal dimension (WFD)	The multiresolution decomposition of the EEG signal is provided by WT analysis. Time-based features such as fractal dimension (FD) computed the complexity of the wavelet coefficients.
Integration of the features	The integration of features is believed to improve the overall classification accuracy.

decomposing the EEG data. This criterion works because the wavelet basis function can provide high values of EPER which best suits the shape of the EEG signals under observation.

9.1.1.3 *Computing Wavelet-Based Coefficients*

Fig. 9.2 shows EEG signal decomposition into detailed and approximate coefficients with the fourth level of decomposition. At each level, the detailed and approximate coefficients were described with labels D1 and A1, respectively. Similarly, each level has its own detailed and approximate coefficients, for example, at level 2—D2 and A2, level 3—D3 and A3, and level 4—D4 and A4. At each level, the EEG signal was high-pass and low-pass filtered, represented as the detailed and approximate coefficients.

Table 9.2 shows wavelet coefficients at distinct levels corresponding to specific bands of frequencies.

Theoretically, wavelet decomposition is essentially a convolution of EEG signal with different scaled and dilated versions of a selected

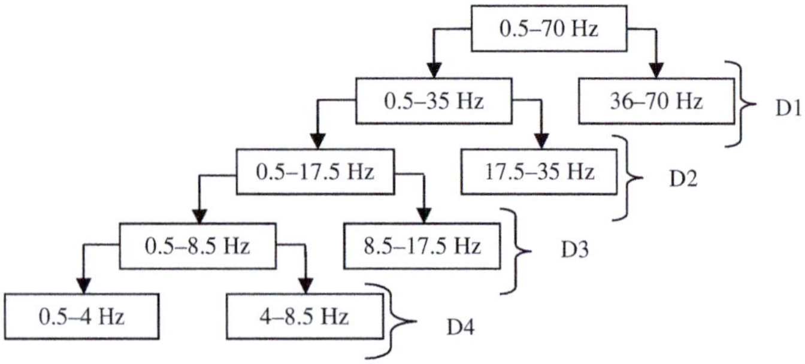

FIGURE 9.2 A multiple-level decomposition of EEG signal.

TABLE 9.2 Wavelet-Based EEG Frequency Bands

Wavelet Coefficients	EEG Frequency Bands	Frequency Ranges (Hz)
D1	Gamma	36–70
D2	Higher beta and low gamma	17.5–35
D3	Alpha and low beta	8.5–17.5
D4	Theta	4–8.5
A4	Delta band	0.5–4

wavelet window function. Each level represents a different scale of the function. In this study, the Daubechies D4 wavelet window was used for the decomposition of the EEG data.[17] The selection of this window function was based on the energy to PE criterion as described in the previous section. According to the theory, wavelet function is defined at scale "a" (dilation parameter) and at location "b" (translation parameter) by the mathematical expression Eq. (9.2):

$$\psi_{a,b}(t) = \frac{1}{\sqrt{a}}\psi\left(\frac{t-b}{a}\right) \tag{9.2}$$

As mentioned in Eq. (9.2), by choosing the orthonormal wavelet basis$\psi_{m,n}(t)$, the original EEG signal can be reconstructed in terms of wavelet coefficients, $T_{m,n}$, using the inverse discrete WT [Eq. (9.3)]:

$$x(t) = \sum_{m=-\infty}^{\infty}\sum_{n=-\infty}^{\infty} T_{m,n}\psi_{m,n}(t) \tag{9.3}$$

reconstruction required the summation of all integers' m and n.

In this chapter, WT analysis was performed [Eq. (9.4)]:

$$\mathrm{DWT}(j,k) = \frac{1}{\sqrt{|2^j|}} \int_{-\infty}^{\infty} x(t)\psi\left(\frac{t-2^j k}{2^j}\right) dt \tag{9.4}$$

where "a" and "b" were replaced by 2^j and $k2^j$, respectively. An efficient way for implementing this scheme was invented by passing the signal through a series of low-pass (LP) and high-pass (HP) filter pairs named as quadrature mirror filters.[18]

9.1.1.4 Computing Wavelet-Based Signal Energy

In this chapter, the total energy contained in EEG signal, $x(t)$, was defined as its integrated squared magnitude [Eq. (9.5)]:

$$E = \int_{-\infty}^{\infty} |x(t)|^2 dt = \|x(t)\|^2 \tag{9.5}$$

The relative contribution of signal energy contained at a specific "a" scale and "b" location is given by the wavelet energy density function [Eq. (9.6)]:

$$E(a,b) = |T(a,b)|^2 \tag{9.6}$$

The scalogram is integrated across "a" and "b" to recover the total EEG signal energy using the admissibility constant, C_g [Eq. (9.7)]:

$$E = \frac{1}{C_g} \int_{-\infty}^{\infty} \int_{0}^{\infty} |T(a,b)|^2 \frac{da}{a^2} db \tag{9.7}$$

The value of admissibility constant depends on the chosen wavelet, for example it is equal to *pi* for the Daubechies wavelet.

9.1.1.5 Computing Wavelet-Based Sample Entropy

In this chapter, WSE was employed to compute the complexity of EEG signal recorded during EC and EO conditions. SE was first developed by Richman and Moorman,[13] and can be computed based on the procedure provided by steps one to six.[19]

The EEG data were decomposed with WT analysis into different wavelet coefficients. The wavelet coefficients were presented as N data points [Eq. (9.8)]:

$$x(1), x(2), \ldots, x(N) \tag{9.8}$$

1. Formulate m dimensional vectors X consecutively from the EEG wavelet coefficients achieved in Eq. (3.17), starting with the ith point [Eq. (9.9)]:

$$X(i) = [x(i), x(i+1), \ldots, x(i+m-1)] \tag{9.9}$$

where $i, j = 1, 2, \ldots, N-m+1, j \neq i$.

2. Next, compute the distance between two consecutive m dimensional vectors as derived in Eq. (9.9) and shown in Eq. (9.10):

$$d[X(i), X(j)] = \max_{k=0,1,\ldots m-1} [|x(i+k) - x(j+k)|] \tag{9.10}$$

3. Given a tolerance window r for every ith value, calculate the distance $d[X(i)$ and $X(j)]$, count the number of this distance, which is less than or equal to r and denoted as $B_i(r)$. Then calculate the ratio of this number to $N-m-1$, indicated in (9.11), $B_i^m(r)$:

$$B_i^m(r) = \frac{1}{N-m-1} B_i \tag{9.11}$$

4. Calculate the average value of $B_i^m(r)$ [Eq. (9.12)]:

$$B^m(r) = \frac{1}{N-m+1} \sum_{i=1}^{N-m+1} B_i^m(r) \tag{9.12}$$

5. Increase the vector dimension from m to $m+1$ and repeat Steps 1–4 and calculate $B_r^{m+1}(r)$:

$$B^{m+1}(r) = \frac{1}{N-m+1} \sum_{i=1}^{N-m+1} B_i^{m+1}(r) \tag{9.13}$$

Theoretically, from Eqs. (8.12) and (8.13), the SE can be defined in Eq. (9.14):

$$\mathrm{SE}(m, r, N) = \lim_{N \to \infty} \left\{ -\ln \left[\frac{B^{m+1}(r)}{B^m(r)} \right] \right\} \tag{9.14}$$

6. When N has a finite value, Eq. (9.15) represents the final formula:

$$\mathrm{SE}(m, r, N) = -\ln \left[\frac{B^{m+1}(r)}{B^m(r)} \right] \tag{9.15}$$

In this chapter, the SE has three parameters: a run length m, a tolerance window r, and N which represents the number of data points. Parameter values such as $m=1$ and $r=0.2$ SD were used as suggested by.[13] SD was the standard deviation of the original EEG data $X(i)$.

The same procedure was adopted to compute the WSE for the detection of treatment response and nonresponse. The WSE was computed for each scalp channel and the numerical values described the complexity of the EEG signal at a scalp location. A high (e.g., approx. 1) WSE value indicated more brain complexity (activation), while a low (e.g., approx. 0) WSE value reflected lower complexity (activation), respectively. In this study, it was hypothesized that abnormal brain activity due to MDD treatment response may show different values of WSE as compared with treatment nonresponse. The computation of SE resulted in a signal numerical value for each scalp location or EEG channel. In this study, the EEG data were computed for EC and EO conditions, including the 19 scalp locations, which resulted in 38 numerical values.

9.1.1.6 Computing Wavelet-Based Composite Permutation Entropy Index

Computing WCPEI is a nonlinear method. It is directly proportional to the complexity of a given time series dataset. It can be achieved through the following four steps:[20]

1. The wavelet coefficients achieved by decomposing the EEG signal using WT analysis were fragmented into a sequence of motifs. As shown in Fig. 9.3, the motifs are shapes of six distinct types. These shapes can explain most of the changes in an EEG signal.
2. Identification of each motif is limited to only one of the possible six types.
3. The number of motifs of each type in the signal were counted to obtain the probability of occurrence of each motif.
4. The Shannon uncertainty formula [Eq. (9.16)] was used to calculate PE. It was based on the normalized probability distribution of the motifs.

$$PE = -\frac{\sum p_i \times \ln(p_i)}{\ln(\text{number of motifs})} \tag{9.16}$$

For the calculation of CPEI, two PEs with parameters, noise threshold (tie) and lag values (τ), were added. The mathematical description for CPEI used in the analysis is:

$$CPEI = \frac{\sum p_i \times \ln(p_i)_{\text{tie}<0.5,\tau=1} + \sum p_i \times \ln(p_i)_{\text{tie}<0.5,\tau=2}}{\ln(49)} \tag{9.17}$$

where noise threshold (tie < 0.5 uV) and τ equals either 1 or 2. In this study, the same procedure was adopted to compute the WCPEI for both study groups, that is, treatment responders and treatment nonresponders. The WCPEI was computed for each scalp channel and numerical values described the complexity of the EEG signal at each scalp location.

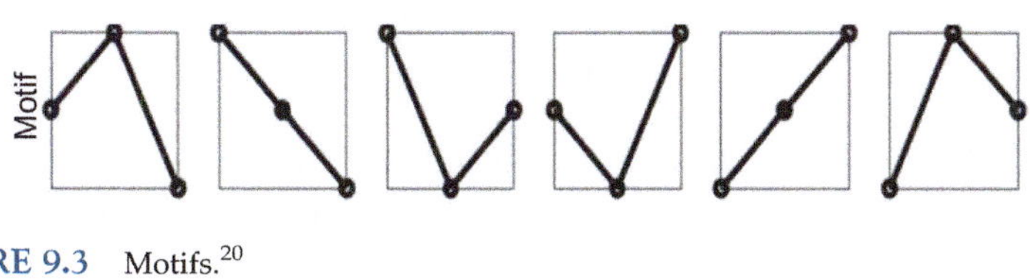

FIGURE 9.3 Motifs.[20]

High (e.g., approx. 1) WCPEI values implicated more brain complexity (activation) while low (e.g., approx. 0) WCPEI values reflected lower complexity (activation), accordingly. In this study, it was hypothesized that treatment response may show different values of WCPEI as compared with treatment nonresponse. The computation of WCPEI resulted in a signal numerical value for each scalp location or EEG channel. In this study, the EEG data were computed for EC and EO conditions, including 19 scalp locations, which resulted in 38 numerical values.

9.1.1.7 Computing Wavelet-Based Fractal Dimension

WFD calculates the complexity and irregularity of a time-based signal. Its value ranges from $D_f = 1$ for a simple curve to $D_f = 2$ for a curve which nearly fills out the whole plain [Eq. (9.18)].

$$L(k) = k^{-D_f} \tag{9.18}$$

The algorithm constructs k new time series based on the wavelet coefficients that were achieved by decomposing the given EEG data: $X(1)$, $X(2),\ldots, X(N)$ [Eq. (9.19)].

$$X_m^k{:}X(m),\ldots, X\left(m + \text{int}\left(\frac{N-m}{k}\right) \times k\right) \tag{9.19}$$

where $m = 1, 2,\ldots,k$, N is the total number of samples, k is interval time, m is initial time, int(r) represents the integer part of a real number r. Eq. (9.20) describes the mathematical formulation for the length $L_m(k)$ of each curve X_m^k:

$$L_m(k) = \frac{1}{k}\left|\sum_{i=1}^{M} |X(m + i \times k) - X(m + (i-1) \times k)|\right|\left(\frac{N-1}{M \times k}\right) \tag{9.20}$$

where

$$M = \text{int}\left(\frac{N-m}{k}\right) \tag{9.21}$$

$L_m(k)$ represents the sum of normalized absolute values of difference in ordinates of pair of points distant k (with initial point m). The curve length $L(k)$ for time interval k is calculated as the mean of k values, that is, $L_m(k)$ where $m = 1, 2,\ldots, k$ [Eq. (9.22)]:

$$L(k) = \left(\frac{\sum_{m=1}^{k} L_m(k)}{k}\right) \tag{9.22}$$

where $L_m(k)$ represents the sum of normalized absolute values of difference in ordinates of pair of points distant k (with initial point m). The value of FD D_f is calculated by a least square linear best-fitting procedure. It is equivalent to the slope coefficient of the linear regression of the log/log graph [Eq. 9.22].

The same procedure was adopted to compute the WFD for both study groups, that is, treatment response and treatment nonresponse. The WFD was computed for each scalp channel and numerical values described the complexity of the EEG signal at each scalp location. High (e.g., approx. 1) WFD values implicated more brain complexity (activation) while low (e.g., approx. 0) WFD values reflected lower complexity (activation), respectively. In this study, it was hypothesized that treatment response may show different values of WFD as compared with treatment nonresponse. The computation of WFD resulted in a signal numerical value for each scalp location or EEG channel. In this study, the EEG data were computed for EC and EO conditions, including 19 scalp locations, which resulted in 38 numerical values.

In this paragraph, a brief description is provided regarding the feature selection method employed in this study. A detailed description of the method is provided in Section 7.5 (Feature Selection). In this study, a rank-based feature selection method was employed to find the most significant features in the *EEG data matrix*.[21] According to the method, each feature was assigned a weight value corresponding to the ability of the feature to classify data points into corresponding target classes. The weight values were computed according to the area under curve (AUC) of the receiver operating characteristic (ROC) computed for a feature. A higher weight value was assigned to a feature with a larger AUC than a feature with a smaller AUC. The weight value could be any number between 0 and 1, indicating a bad to good classification ability. The weight values allowed for the most significant features to be listed in descending order. According to this method, the most expressive features could be identified by selecting the top-listed features only.

9.1.1.8 Integration of Features

The integration of these features was performed based on the most noteworthy features identified individually from *WCPEI, WSE,* and *WFD* (e.g., *features from WCPEI, WSE, and WFD*). In addition to these features, the energy of the EEG signal was computed based on wavelet coefficients. Because of the feature extraction, different EEG data matrices were constructed and subjected as input data to the ML scheme (*ITMS-treatment selection*).

9.1.2 Finalizing the Electroencephalography Data Matrix

Table 9.3 provides the dimensions of each *EEG data matrix*. Because of the feature extraction, a substantial number (N_c) of candidate features

TABLE 9.3 EEG Data Matrices and Dimensions

EEG Data Matrix	Matrix Dimension (per band*)	
EEG absolute power delta: the "64" corresponds to 34 MDD patients and 30 healthy controls. The "38" corresponds to 19 scalp locations for EC and 19 for EO.	64 × 38*	ITMS diagnosis
EEG alpha interhemispheric asymmetry: the "64" corresponds to 34 MDD patients and 30 healthy controls. The "128" corresponds to 8 scalp locations on the left and 8 on the right hemisphere, i.e., 8 × 8 = 64, and for each EC and EO, i.e., 64 × 2 = 128.	64 × 128	
Synchronization likelihood: the "64" corresponds to 34 MDD patients and 30 healthy controls. The "171" corresponds to all possible combinations of scalp sensors.	64 × 171	
Integration of features (ITMS diagnosis)	64 × 337	
Wavelet-based energy (WE): the "64" corresponds to 34 MDD patients and 30 healthy controls. The "38" corresponds to 19 scalp locations for EC and 19 for EO.	34 × 38*	ITMS-treatment selection
Wavelet-based CPEI (WCPEI): the "64" corresponds to 34 MDD patients and 30 healthy controls. The "38" corresponds to 19 scalp locations for EC and 19 for EO.	34 × 38*	
Wavelet-based FD (WFD): the "34" corresponds to 34 MDD patients. The "38" corresponds to 19 scalp locations for EC and 19 for EO.	34 × 38*	
Wavelet-based SE (WSE): the "34" corresponds to 34 MDD patients. The "38" corresponds to 19 scalp locations for EC and 19 for EO.	34 × 38*	
Integration of features (ITMS-treatment selection)	34 × 152	

**The corresponds to only one EEG band. It could be delta or theta or alpha or beta or gamma.*

were computed and arranged in a matrix columnwise, each column was denoted as x_i, where $i = 1 \ldots N_c$. In this study, the number of features was in the thousands indicating its high dimensionality. In addition, the rows of the matrix represent MDD patients with physiological conditions provided for each patient. In this study, this feature space matrix was termed as *EEG data matrix*. The feature space denoted by $L = [(x_i, y_i), i = 1 \ldots N_c]$ included both the feature space matrix and the corresponding output class labels, that is, $y =$ [MDD, Controls] or

$y = [R, NR]$. Finally, it was observed that the resulting matrix (L) was rectangular and included high-dimensional datasets (the number of columns) with 64 data points (the number of rows).

The diagnosis and treatment outcome prediction involved two different methods. Regarding the diagnosis, EEG absolute power and alpha asymmetry were considered as vulnerability markers for diagnosing depression. These features showed association (statistically significant) with the MDD patients as compared to the healthy controls. This finding suggests that averaging the PSD (EEG absolute values) to compute absolute power and asymmetry could be useful features for discriminating between MDD patients and healthy controls.

On the other hand, discrimination within the MDD patients is more challenging than discriminating between the MDD patients and healthy controls. For example, it required micro-level details of the EEG data, as evident from a previous study,[22] where the coherence was computed at high frequency resolution such as at each Hz. The study successfully discriminated the treatment respondents from the nonrespondents among a group of MDD patients. Moreover, as EEG has high temporal resolution, time-based quantities such as CPEI, SE, and FD have been reported as promising measures to quantify brain behavior. For example, CPEI has been considered as a measure of GABAergic hypnotic drug effects. SE represents the signal complexity of the bioelectric processes in the brain. For example, SE values for patients with cerebral injuries were found lower than normal individuals. FD provides the ability to distinguish different pathophysiological states by directly examining EEG in time domain.

Based on these findings, in this study it was hypothesized that methods such as the absolute power of different bands and asymmetry are more suitable for discriminating MDD patients from healthy controls, and hence useful for EEG-based diagnosis of depression. On the other hand, discrimination within the MDD patients group requires a deeper level of analysis such as the multiresolution decomposition of the EEG signal, and hence suitable for EEG-based discrimination of antidepressant treatment response and nonresponse.

9.2 TREATMENT RESPONDENTS VERSUS NONRESPONDENTS

The differences between the R and NR are studied from different perspectives such as through sLORETA analysis which provides differences between the two groups. In addition, topographic maps provide a unique perspective of looking at the differences between R and NR.

9.2.1 sLORETA Analysis

Based on infinitely referenced EEG data, sLORETA analysis was applied for EC and EO conditions with gender stratified groups. The gender stratification was performed to determine brain regions that might not be revealed in the case of a combined group analysis. The significance of the differences between responders and nonresponders was determined by performing a student's *t*-test.

Table 9.4 shows sLORETA maps of activations to localize brain areas that are statistically significantly different between R and NR during EC and EO conditions. According to the table, during EC condition, the frontal brain area has shown high activations while comparing responders with nonresponders. In EO condition, two brain regions were found to be significantly different, that is, the frontal and occipital areas.

According to sLORETA analysis (Table 9.4), the frontal and occipital brain regions are localized and show statistically significant difference between the treatment respondents and nonrespondents. According to the literature, the differences of the activations between groups suggest the importance of utilizing a classification model that is capable of learning these patterns between R and NR. Brain source localization (BSL) techniques such as LORETA and its variants, including sLORETA

TABLE 9.4 sLORETA Maps for Respondents Versus Nonrespondents

Student's *t*-test-based Differences Between R and NR	Brain Areas
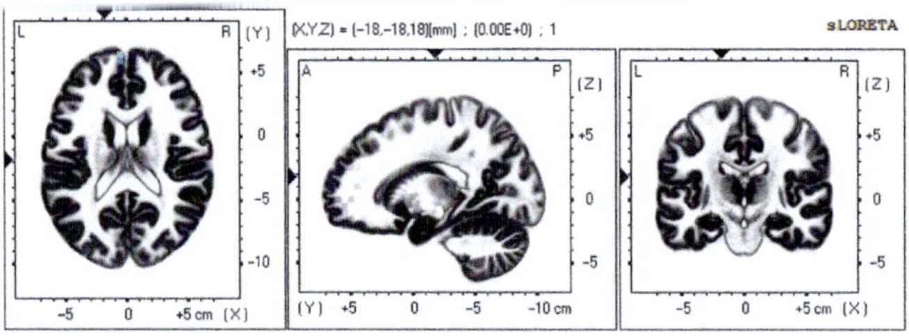 (Eyes closed condition)	Frontal BA = 11
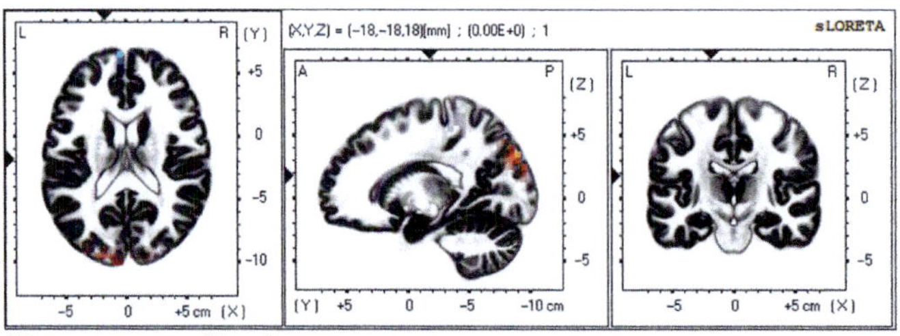 (Eyes open condition)	Frontal BA = 11 Occipital BA = 19

and standardized shrinking LORETA-FOCUSS (ssLOFO), are used to localize brain regions. Studies based on BSL techniques have identified activations of brain areas that are associated with treatment outcome either as treatment responders or nonresponders. The BSL was applied in studies,[23,24] where 18 MDD patients were treated with Nortriptyline for 16 weeks. In addition, the patients were treated with Citalopram and Reboxitine, respectively. Both studies resulted in increased pretreatment resting delta activity in the rostral anterior cingulate cortex (rACC). The activity was associated with treatment response. In another study, higher theta activity in the rACC and orbitofrontal cortex was found to be correlated with response to medication.[25]

9.2.2 Topographic Maps

Figs. 9.4 and 9.5 show wavelet-based topographic maps[26] for MDD versus healthy controls and responders versus nonresponders, respectively. Fig. 9.4 shows the difference between participants based on the Wilcoxon-based rank sum as exhibited by the *P*-values. During EC, brain regions such as the frontal, left, and right temporal showed significant differences. In addition, some other areas such as left central, parietal, and occipital also depicted significant differences. During EO, subjects exhibited differences in the right frontal and temporal areas on both sides. In addition, right sided occipital and parietal also showed significant differences. During EC and EO conditions, frontal and temporal areas were commonly observed as significantly different between the two groups, which is in accordance with the literature.[27]

In Fig. 9.5, during EC, statistically significant differences in the right frontal, left temporal, and right parietal regions were observed. During EO, the subjects exhibited differences in the left temporal area. During EO, in addition to the frontal and temporal areas, the sensor at the

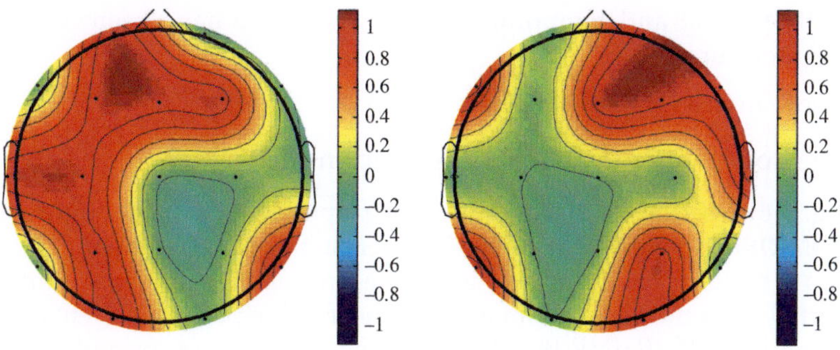

FIGURE 9.4 Topographic maps for MDD versus controls during EC (left) and EO (right).

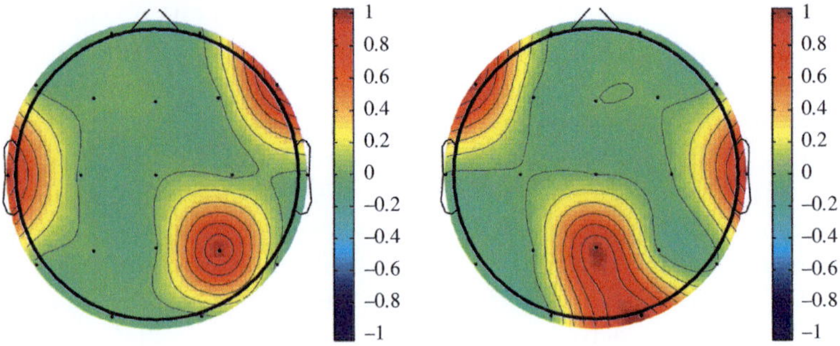

FIGURE 9.5 Topographic maps for R versus NR during EC (left) and EO (right).

central and parietal regions showed statistical differences. Again, it was observed that the frontal and temporal regions were common between EC and EO conditions.

9.2.3 Ranking Features Based on Receiver Operating Characteristic Criterion

Table 9.5 shows the most significant features according to ROC criterion during the feature selection process. The features were arranged in descending order according to their individual z-values. Higher z-values directly correspond to the ability of a feature to discriminate between the respondents and nonrespondents. Therefore, the z-values may vary between 0 and 0.5, indicating bad to good classification ability, respectively. Among the 15 most significant features, 9 were associated with the frontal lobe and 4 were found associated with the temporal region. The parietal and central areas implicated one and two features, respectively. In summary, the frontal and temporal areas have implicated highly significant features (frontal = 9, temporal = 4, parietal = 1, and central = 2). Abnormalities associated with the frontal and temporal areas during MDD have been reported in the literature as well.[27]

9.2.4 Low-Dimensional Representation

Fig. 9.6 describes the distribution of responders (R) and nonresponders (NR) on a low-dimensional (2-dimensional) representation. Fig. 9.6 shows a scatter plot of Mt = 64 available pretreatment training samples projected onto the first two major nonlinear principal components only. As shown in the figure, the shapes of the two clusters provide a data visualization in terms of responders and nonresponders.

TABLE 9.5 A List of Discriminating Feature Indices of ITMS-Treatment Selection

No.	EEG Electrodes	Frequency Band	Absolute *z*-values	*P*-values
1	Fp2	Delta	0.3024	.016
2	C3	Theta	0.2886	.022
3	F7	Delta	0.2794	.013
4	F3	Delta	0.2794	.022
5	F7	Theta	0.2739	.016
6	T4	Theta	0.2711	.022
7	F8	Theta	0.2711	.008
8	T4	Delta	0.2711	.010
9	F3	Theta	0.2665	.002
10	Fz	Delta	0.2656	.045
11	F4	Delta	0.2638	.0021
12	C4	Delta	0.2601	.015
13	F8	Delta	0.2574	.021
14	T4	Theta	0.2555	.030
15	P3	Delta	0.2555	.001

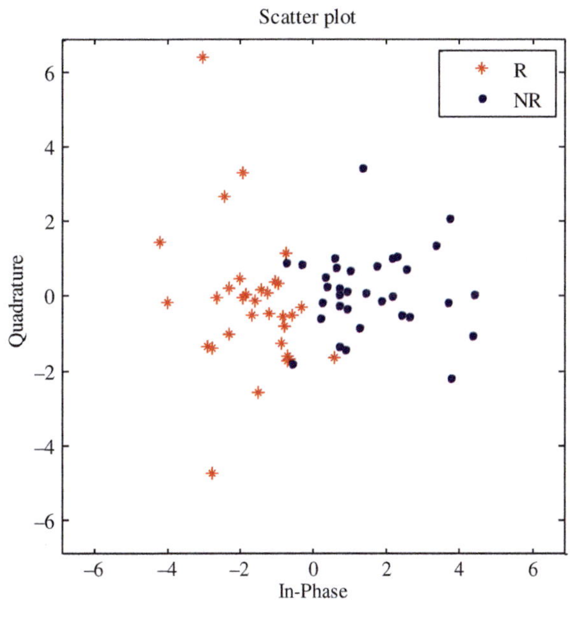

FIGURE 9.6 A 2D representation of R versus NR using PCA.

Fig. 9.6 shows low-dimensional representations based on kernel-based principal component analysis (KPCA). The KPCA can be used to visualize the clustering behavior of the multidimensional feature space. This figure was generated using the KPCA method with a Gaussian kernel using Nr = 15 selected features. The x-axis represents the first PCA component and the y-axis corresponds to the second PCA component. This example lends credibility to the idea that it is possible to select a set of features from background EEG which are indicative of response. An advantage of the low-dimensional representation (Fig. 8.6) is that it visually confirms that the classes do indeed cluster in distinct separable regions in the feature space, indicating that prediction is feasible.

9.2.5 Classification Results (ITMS-Treatment Selection)

Table 9.6 provides SVM classification results for discriminating between treatment respondents and treatment nonrespondents. The results are resimulated using data collected from this work. According to the table, the wavelet-based CPEI features performed better than the wavelet-based energy, FD, and SE features (accuracy = 80.1%, sensitivity = 81.6%, specificity = 77.8%). The integration of all the wavelet-based features (*ITMS-treatment selection*) resulted in accuracy = 89.1%, sensitivity = 91%, and specificity = 88.7%.

Table 9.7 provides a comparison between the ML method and state-of-the-art methods (as mentioned in the Chapter 5, Sections 5.2 and 5.3) alpha, theta power, alpha asymmetry, ATR index, theta cordance, coherence, and P300 intensities. The results are resimulated using data collected from this work. The table presents results from the logistic regression classifier with the IR-referenced EEG data only. The number of features reported here show maximum classification accuracies with the given feature sets. Results shown in the table signify better performance of the wavelet-based method than the existing state-of-the-art methods. It can be observed that the second-best accuracy is provided

TABLE 9.6 SVM Classification for R Versus NR

EEG-Based Features	Accuracy (%)	Sensitivity (%)	Specificity (%)
Wavelet-based energy	77.2	78.3	76.4
Wavelet-based CPEI	80.1	81.6	77.8
Wavelet-based SE	74.44	71.66	76.6
Wavelet-based FD	77.3	80	75
ITMS-treatment selection	89.1	91	88.7

TABLE 9.7 Comparison Between ITMS-Treatment Selection and Literature

EEG-Based Methods	Accuracy (%)	Sensitivity (%)	Specificity (%)
Alpha power	62.2	64.7	60
Theta power	58.71	64.7	52.14
Alpha asymmetry	65	62.38	68.5
ATR index	61.68	70	54
Theta cordance	70.7	75.7	65.7
Coherence	76.3	81.9	69.5
P300 intensities	74.16	70	75
ITMS-treatment selection	**89.1**	**91**	**88.7**

TABLE 9.8 SVM Classification (R vs NR) for EEG Data

	Classification Performance		
EEG Features	**Accuracy (%)**	**Sensitivity (%)**	**Specificity (%)**
Wavelets	**89.1**	**91**	**88.7**
STFT	61.66	52.5	70
EMD	60.4	52.5	70
Wavelet, STFT, and EMD	85.33	84.5	86.5

by coherence, that is, 76.3%. However, the associated specificities are low, which leads to the conclusion that these features are not suitable for clinical applications. Finally, integration of all the wavelet-based features (*ITMS-treatment selection*) have resulted in accuracy = 89.1%, sensitivity = 91%, and specificity = 88.7%.

Table 9.8 provides a comparison of the ML method with STFT and EMD techniques. The EEG features computed with WT analysis showed the highest classification efficiencies (accuracy = 89.1%) among other EEG features. On the other hand, STFT- and EMD-based EEG feature extraction showed a lower performance than WT analysis. An integration of the features including WT analysis, EMD, and STFT resulted in an accuracy of 85.3%.

The ML techniques utilized were recommended by Ahmad-khudayari and fellow researchers, for example, utilization of ML techniques are recommended to classify study participants based on pretreatment EEG data.[28] However, ML methods discussed in this chapter include logistic regression classification, support vector machine, and Naïve Bayesian classification, which were not employed by previous studies. Other parts of the scheme such as feature extraction, selection, and validation are also different from previous studies. However, the wavelet-based method offers a higher efficiency (accuracy, sensitivity, and specificity) involving wavelet features extracted from EEG. In addition, wavelet features from the frontal and temporal areas are found to be significant. This finding is in accordance with various research studies related to MDD.[29]

Regarding the question of clinical applicability, the wavelet-based method provides a faster evaluation of treatment outcomes as it utilizes pretreatment EEG data. This contrasts with methods that require data from the first week as well, for example, the ATR index.[30] Furthermore, the wavelet-based method provides higher values of specificities, which favor its clinical utility. Biomarkers such as theta cordance and ERP-based techniques (P300 and LDEAP) resulted in low specificities.[31,32]

The selected antidepressants have similar mechanisms of action and are categorized under the same class of antidepressant, that is, SSRIs. Therefore, the wavelet-based method is applicable to these kinds of antidepressants only. Other classes of antidepressants such as SNRIs or tricyclics have different mechanisms of actions, and this method requires further investigation for applicability to these kinds of treatments. Therefore, caution should be adopted when generalizing results for antidepressants, though they may come under the category of SSRIs.

9.3 DISCUSSION

In this chapter, an improved feature selection and classification system termed as *ITMS-treatment selection* is presented, which utilizes EEG wavelet-based features as biomarkers for antidepressant treatment selection involving SSRIs. *ITMS-treatment selection* shows a higher efficiency than published methods. In addition, treatment efficacy assessment is performed on EEG data (week 0) recorded from drug naïve MDD patients. Conventionally, an adequate time frame of 2–4 weeks is required for antidepressant treatment efficacy assessment. Hence, *ITMS-treatment selection* might improve treatment selection by effectively reducing the time frame of 2–4 weeks. In addition, the wavelet-based system may help psychiatrists during treatment selection by providing objective evidence to evaluate the suitability of antidepressants under the SSRIs category.

In this chapter, brain localization techniques help in identifying brain areas such as frontal, temporal, parietal, and occipital. The techniques included (1) the construction of 3D images based on sLORETA analysis; (2) topographical maps generated from a reduced set of wavelet features; and (3) the 15 top-ranked features sorted according to ROC criterion. This finding is in accordance with other research studies related to MDD.[29] Other studies based on structural observations such as MRI, including MDD patients with abnormalities associated with the frontal, temporal, parietal and occipital regions.[29,33–35] However, the main contribution is that our data have replicated these findings with wavelet decomposition.

There is a possibility that our ML-based models are confounded with some outliers other than relevant patterns extracted from brain activities. However, this concern has been ruled out by (1) properly adopting artifact removal techniques; (2) standardizing preprocessed data based on z-scores; (3) plotting the low-dimensional representation of the feature space. This helps in identifying outliers which may disturb interpretations and conclusions; (4) during the classifier testing and training, selecting random data points so that each data point in the feature space can be used; and (5) in terms of classification, equally distributing both the responder and nonresponder classes within MDD male and female patients. Based on all these precautions, it may be concluded that the results shown here are unbiased and a true representation of the information from the recorded pretreatment EEG data.

9.4 SUMMARY

This chapter provides details on the ITMS for treating depression. In addition, few limitations of the proposed method have been highlighted. For example, during MDD patient recruitment, it was difficult to recruit patients under a common treatment. As a result, the inclusion of patients was restricted to a single class of antidepressants, that is, SSRIs. Since the pharmaco-EEG profiles of different antidepressants are not clear yet, it is difficult to study medication-specific treatment effects. The proposed ITMS has been validated with relatively small sample size. In addition, the findings are specific to Malaysian populations only. Therefore, in order to generalize the findings, it is necessary to replicate the reported results into larger population. The study patients are required to be in washout for a period of 2 weeks before the first EEG data acquisition. In future studies, the inclusion of psychophysiological characteristics integrated with EEG may improve prediction performance.

References

1. Faust O, Acharya UR, Adeli H, et al. Wavelet-based EEG processing for computer-aided seizure detection and epilepsy diagnosis. *Seizure*. 2015;26:56–64.

2. Hramov AE, Koronovskii AA, Makarov VA, et al. *Wavelet approach to the study of rhythmic neuronal activity. Wavelets in Neuroscience*. Springer; 2015:177–209.
3. Faust O, Ang PCA, Puthankattil SD, et al. Depression diagnosis support system based on EEG signal entropies. *J Mech Med Biol*. 2014;14(03):1–9.
4. Addison PS. *The Illustrated Wavelet Transform Handbook: Introductory Theory and Applications in Science, Engineering, Medicine and Finance*. CRC Press; 2010.
5. Adeli H, Ghosh-Dastidar S, Dadmehr N. A wavelet-chaos methodology for analysis of EEGs and EEG subbands to detect seizure and epilepsy. *IEEE T Bio-Med. Eng*. 2007;54 (2):205–211.
6. Adeli H, Ghosh-Dastidar S, Dadmehr N. A spatio-temporal wavelet-chaos methodology for EEG-based diagnosis of Alzheimer's disease. *Neurosci Lett*. 2008;444 (2):190–194.
7. Kwon JS, Youn T, Jung HY. Right hemisphere abnormalities in major depression: quantitative electroencephalographic findings before and after treatment. *J Affect Disord*. 1996;40(3):169–173.
8. Korb AS, Hunter AM, Cook IA, et al. Rostral anterior cingulate cortex theta current density and response to antidepressants and placebo in major depression. *Clin Neurophysiol*. 2009;120(7):1313–1319.
9. Davidson RJ, Henriques JB. Regional brain function in sadness and depression. In: *The Neuropsychology of Emotion*; 2000:269–297.
10. Herrington JD, Heller W, Mohanty A, et al. Localization of asymmetric brain function in emotion and depression. *Psychophysiology*. 2010;47(3):442–454.
11. Olofsen E, Sleigh J, Dahan A. Permutation entropy of the electroencephalogram: a measure of anaesthetic drug effect. *Br J Anaest*. 2008;101(6):810–821.
12. Dandan Z, Haiyan D, Xinlin D, et al. The combination of amplitude and sample entropy in EEG and its application to assessment of cerebral injuries in piglets. In: *2008 International Conference on BioMedical Engineering and Informatics*. Sanya, China; 2008:525–529.
13. Richman JS, Moorman JR. Physiological time-series analysis using approximate entropy and sample entropy. *Am J Physiol Heart Circ Physiol*. 2000;278(6): H2039–H2049.
14. Jiayi G, Peng Z, Xin Z, et al. Sample entropy analysis of sleep EEG under different stages. In: *2007 IEEE/ICME International Conference on Complex Medical Engineering*. Beijing, China; 2007.
15. Accardo A, Affinito M, Carrozzi M, et al. Use of the fractal dimension for the analysis of electroencephalographic time series. *Biol Cybern*. 1997;77(5):339–350.
16. Higuchi T. Approach to an irregular time series on the basis of the fractal theory. *Physica D*. 1988;31(2):277–283.
17. Kalayci T, Ozdamar O. Wavelet preprocessing for automated neural network detection of EEG spikes. *IEEE Eng Med Biol Mag*. 1995;14(2):160–166.
18. Mallat SG. A theory for multiresolution signal decomposition: the wavelet representation. *IEEE Trans Pattern Anal Mach Intell*. 1989;11(7):674–693.
19. Li X, XU G, Yang S. Sample entropy of EEG based on acupuncturing. In: *The 4th International Conference on Bioinformatics and Biomedical Engineering (iCBBE)*. Tianjin, China; 2010.
20. Xiaoli L, Cui S, Voss LJ. Using permutation entropy to measure the electroencephalographic effects of Sevoflurane. *Anesthesiology*. 2008;109(3):448–456.
21. Mamitsuka H. Selecting features in microarray classification using ROC curves. *Pattern Recogn*. 2006;39(12):2393–2404.
22. Khodayari-Rostamabad A, Reilly HP, Hasey GM, et al. A machine learning approach using EEG data to predict response to SSRI treatment for major depressive disorder. *Clin Neurophysiol*. 2013;124(10):1975–1985.

23. Pizzagalli D, Marqui RDP, Nitschke JB, et al. Anterior cingulate activity as a predictor of degree of treatment response in major depression: evidence from brain electrical tomography analysis. *Am J Psychiatry*. 2001;158:405–415.
24. Mulert C, Juckel G, Brunnmeier M, et al. Rostral anterior cingulate cortex activity in the theta band predicts response to antidepressive medication. *Clin EEG Neurosci*. 2007;38(2):78–81.
25. Korb AS, Hunter AM, Cook IA, et al. Rostral anterior cingulate cortex theta current density and response to antidepressants and placebo in major depression. *Clin Neurophysiol*. 2009;120(7):1313–1319.
26. Plante D, Goldstein M, Landsness E, et al. Topographic and sex-related differences in sleep spindles in major depressive disorder: a high-density EEG investigation. *J Affect Disord*. 2013;146(1):120–125.
27. Price JL, Drevets WC. Neural circuits underlying the pathophysiology of mood disorders. *Trends Cogn Sci*. 2012;16(1):61–71.
28. Khodayari-Rostamabad A, Reilly JP, Hasey GM, et al. A machine learning approach using EEG data to predict response to SSRI treatment for major depressive disorder. *Clin Neurophysiol*. 2013;124(10):1975–1985.
29. Sheline YI. Neuroimaging studies of mood disorder effects on the brain. *Biol Psychiatry*. 2003;54(3):338–352.
30. Leuchter AF, Cook IA, Marangell LB, et al. Comparative effectiveness of biomarkers and clinical indicators for predicting outcomes of SSRI treatment in major depressive disorder: results of the BRITE-MD study. *Psychiatry Res*. 2009;169(2):124–131.
31. Isıntas M, Ak M, Erdem M, et al. Event-related potentials in major depressive disorder: the relationship between P300 and treatment response. *Turk Psikiyatri Derg*. 2012;23(1):33–39.
32. Gallinat J, Bottlender R, Juckel G, et al. The loudness dependency of the auditory evoked N1/P2-component as a predictor of the acute SSRI response in depression. *Psychopharmacology (Berl)*. 2000;148(4):139–143.
33. Koolschijn P, van Haren NE, Lensvelt-Mulders GJ, et al. Brain volume abnormalities in major depressive disorder: a meta-analysis of magnetic resonance imaging studies. *Hum Brain Mapp*. 2009;30(11):3719–3735.
34. Hasler G. Pathophysiology of depression: do we have any solid evidence of interest to clinicians? *World Psychiatry*. 2010;9(3):155–161.
35. Lorenzetti V, Allen NB, Fornito A, et al. Structural brain abnormalities in major depressive disorder: a selective review of recent MRI studies. *J Affect Disord*. 2009;117(1):1–17.

Index

Note: Page numbers followed by "*f*" and "*t*" refer to figures and tables, respectively.

E

F

CPI Antony Rowe
Eastbourne, UK
September 12, 2019